Numerical Exercises in Fire Protection Engineering

Numerical Exercises in Fire Protection Engineering

J C Jones DSc, FIChemE, FRSC
Department of Engineering, University of Aberdeen

Whittles Publishing

Published by
Whittles Publishing,
Dunbeath,
Caithness KW6 6EG,
Scotland, UK

www.whittlespublishing.com

reprinted 2015
ISBN 978-1870325-48-6

Printed and bound in Great Britain by
4edge Ltd., Hockley. www.4edge.co.uk

Dedicated to

Professor Brian F Gray

Preface

I read that a new chemical element has been identified (*Chemical and Engineering News*, November 15th 2004, page 7). It has been assigned the name Roentgenium and the symbol Rg. When men and women first began to conceive that physical reality is divisible into elements they postulated the existence of four: earth, air, fire and water. When I was a chemistry undergraduate 30+ years ago one of our lecturers made the point that dividing what is observable into elements, even on the highly simplified basis of there being only four, was a major intellectual advance which the pioneers in chemistry later made their own.

Though the four-element idea was prevalent by the time of Aristotle (384–322 BC) who himself promoted it, fire must have been thought of as an element at least as recently as the middle of the nineteenth century. In *Jane Eyre*, first published in 1847, the action of Mr Rochester's insane wife in burning down the house to which she had to be confined is recounted. Earlier in the novel, Mrs Rochester attempts to set fire to the room in which her unfortunate husband is sleeping. In the account of Jane Eyre's successful intervention with water to extinguish the fire, the expression "quenched element" is used.

Those who placed fire and water amongst the elements were obviously aware that each has the potential to rob a human being of his or her life. Such realities transcend changes in culture over the generations, and we who live in the early part of the 21st Century are constantly being informed of fatalities through accidental fires. The branch of engineering referred to as Fire Protection Engineering has evolved in order that knowledge in areas including chemistry, physics, structural engineering, mechanical engineering and statistics can be focused on fire safety. The benefits in terms of preservation of life and assets are such that the role of the Fire Protection Engineer in industrial settings and in the operation of such facilities as airports, campuses and shopping malls is well established.

Technology and Applied Science require that different basic scientific topics are brought together and focused on the subject of interest. That is true of this text, in particular thermochemistry and heat transfer. Most importantly, an informed imagination has been applied in the development of realistic numerical problems requiring for an approximate solution only undergraduate-level principles and hand calculation. The word 'approximate' is sometimes used in a facile or dismissive way but that is not the intention here. In many engineering problems, a simplified solution is the first step. Whether the simplified solution can be accepted as the basis for subsequent action or recommendation depends on its bottom line. When a particular occurrence or course of events is being considered, a simple calculation might indicate the total absence of a hazard, in which case there has been sufficient investigation. If the result of the simple calculation is less clear there can be further analysis using more advanced techniques. If the simple calculation indicates very strongly that there *is* a hazard, finer screening will be unnecessary and mitigation or total elimination of the occurrence under consideration is the way to go. In all three cases, a calculation at the level in this book will have been helpful.

Although the division of the book into chapters makes for user-friendliness the reader should be aware of the interrelatedness of the topics. For example, although the topic of Chapter 2 is Flashover, the topic also features in other chapters including Chapter 5: Vehicular fires. In the development of these 90-odd numerical examples, I have drawn on many recent authoritative works and am indebted to two volumes in particular: the latest SFPE Handbook and Vyto Babrauskas' *Ignition Handbook.*

I am confident that the text will be of interest and benefit to readers at a range of levels. In a review of one of my recent books the phrase 'confidence building' was used, which I hope can also be applied to *Numerical Exercises in Fire Protection Engineering.* I also hope that there will be readers whose confidence is already well and truly built who will benefit from this text.

J C Jones

Aberdeen

Contents

Notation List

Symbol	*Definition*	*Units*
A	area of radiating surface	m^2
c	heat capacity	$J\ kg^{-1}\ K^{-1}$
c_g	speed of sound in unburnt gas	$m\ s^{-1}$
CHF	Critical Heat Flux	$kW\ m^{-2}$
d	nozzle diameter	m
D	diameter;	m
	diffusion coefficient	$m^2\ s^{-1}$
e	wall thickness	m
E	activation energy	$J\ mol^{-1}$
f	design stress	$N\ mm^{-2}$
F	flow of persons;	$s^{-1}\ m^{-1}$
	view factor	–
FPI	Fire Propagation Index	–
Gr	Grashof number	–
h	convection coefficient	$W\ m^{-2}\ K^{-1}$
h_R	radiation heat transfer coefficient	$W\ m^{-2}\ K^{-1}$
H	height	m
ΔH_{rad}	heat transferred by radiation per amount of fuel burnt	$MJ\ kg^{-1}$
ΔH_{con}	heat transferred by convection per amount of fuel burnt	$MJ\ kg^{-1}$
ΔH_{ch}	chemical heat of combustion	$MJ\ kg^{-1}$
I	impulse	Pa s
k	thermal conductivity	$W\ m^{-1}\ K^{-1}$
k_f	thermal conductivity of air at film temperature	$W\ m^{-1}\ K^{-1}$
K_L	kinetic energy correction factor	–
L	heat transfer rate	W
	horizontal dimension	m
Le	Lewis number	–
LFL	Lower Flammability Limit	per cent
m	mass flow rate;	$kg\ s^{-1}$
	detector mass	kg
m'	fluid flow rate	$kg\ s^{-1}$
M	Mach number;	–
	moisture content	per cent
Nu	Nusselt number	–
P	pressure	Pa
ΔP	head pressure	Pa

Pr	Prandtl number	–
q	rate of heat transfer	W
q'	heat flux	$W\ m^{-2}$
Q	heat–release rate	W
R	gas constant	$J\ K^{-1}\ mol^{-1}$
R	heat release	W
Ra	Rayleigh number	–
RVP	Reid Vapour Pressure	kPa
S	speed of movement of person	$m\ s^{-1}$
S_o	speed of movement of person in absence of restrictions	$m\ s^{-1}$
t_{ig}	ignition time	s
T	temperature	°C or K
T_o	surrounding temperature	°C or K
T_f	film temperature	°C or K
ΔT	difference between surface and air temperature	°C or K
TRP	Thermal Response Parameter	$kW\ s^{0.5}\ m^{-2}$
u	velocity	$m\ s^{-1}$
UFL	Upper Flammability Limit	per cent
V	volume	m^3
w	weight	kg
W	mass of leaked fuel	kg
x	vertical dimension	m
α	thermal diffusivity	$m^2\ s^{-1}$
β	compressibility factor; volume coefficient of expansion	K^{-1}; K^{-1}
χ	overall heat transfer coefficient	$W\ m^{-2}\ K^{-1}$
δ	density of occupancy	m^{-2}
δ_p	thermal penetration depth	m
ε	emissivity	–
γ	ratio of principal specific heats	
μ	dynamic viscosity	$kg\ m^{-1}\ s^{-1}$
ν	kinematic viscosity	$m^2\ s^{-1}$
ρ	density	$kg\ m^{-3}$
σ	Stefan Boltzmann constant	$W\ m^{-2}\ K^{-4}$
τ	detector time constant	s

Chapter 1:
Fire loads

Introduction

The fire load of a building is the average weight of combustibles per square foot of floor area. In this context a combustible material is one with a calorific value of 16–19 MJ kg^{-1}, the range of values for seasoned wood. As we have seen, some materials have a higher calorific value than this, and in building practice this is allowed for by adjusting the weights (see Exercise 1.1 below). Schools and residential buildings have fire loadings typically in the 7–8 lb ft^{-2} range, while industrial premises may have 25 lb ft^{-2} and storage premises 30 lb ft^{-2}. High-hazard premises are those whose business necessitates the presence of large quantities of materials with high calorific values, e.g. solvents, tyres, granular plastics. Special fire safety requirements apply to such.

> 1.1 Suppose that a room contains 100 lb of wood and 350 lb of polymethylmethacrylate (PMMA) a.k.a. perspex (calorific value 26 MJ kg^{-1}), occupying an area of 40 ft^2. What is the fire load?

Solution

That due to the wood = 100/40 = 2.5 lb ft^{-2}

That due to the PMMA = $[350 \times (26/17)]/40 = 13.4$ lb ft^{-2}

Total = 16 lb ft^{-2} (to nearest whole number)

Note that the contribution from the PMMA has to be scaled as the material has a significantly higher calorific value than seasoned wood which is the benchmark material. Imperial units continue to be used to express fire loads. We shall determine the conversion factor to SI as part of the next question.

1.2 Express a fire load of 30 lb ft^{-2} in units of kg m^{-2}.
Note that: 2.205 lb = 1 kg, 2.54 cm = 1 inch.
Newsprint (calorific value 15 MJ kg^{-1}) is stored in an otherwise empty cellar area with stone floors and walls. The floor area = 15m × 15m square. How much newsprint can be stored in the cellar without exceeding 30 lb ft^{-2} fire load?

Solution

1 ft^2 = 0.093 m^2

30 lb ft^{-2} = 30 lb per 0.093 m^2 or 326 lb m^{-2} or 146 kg m^{-2}

Let the weight of newsprint = w, then:

$146 = w \times (15/17)/225$ kg m^{-2}

$\Rightarrow w = 37230$ kg (37 tonne)

1.3 The same cellar as in the above question is used to store 15 drums each containing 1 m^3 of kerosene. The density of the kerosene is 750 kg m^{-3} and its calorific value 45 MJ kg^{-1}. What is the fire load in kg m^{-2}?

Solution

Fire load = $[(45/17) \times 15 \times 750]/225$ kg m^{-2} = 132 kg m^{-2} (27 lb ft^{-2})

1.4 In the room in question 1.1 containing 100 lb of wood, the 350 lb of PMMA are replaced with an equivalent weight of polystyrene, calorific value 39.5 MJ kg^{-1}. Calculate the fire load using a value of 40 ft^2 for the area as previously.

Solution

That due to the wood = 100/40 = 2.5 lb ft^{-2}

That due to the polystyrene = $[350 \times (39.5/17)]/40$ = 20.3 lb ft^{-2}

Total = 23 lb ft^{-2}

Note

The questions do not specify whether the PMMA or polystyrene is simply being stored or whether it comprises part of the room structure or fittings.

In the latter case, to go from polystyrene to PMMA (that is, in the opposite direction from that in the questions) would be an example of enhanced safety by replacing a hazardous material with a less hazardous one, a principle frequently applied in safety engineering. The substitution of PMMA for polystyrene would reduce the fire load by about a third.

Chapter 2:
Flashover

Introduction

Flashover is the transition from a localised fire to one involving the entire enclosure in which the fire is taking place, and is characterised by a very sudden increase in the heat–release rate and in the temperature. Useful flashover models have been developed which assume single pre- and post-flashover temperatures. These models invoke the classical Semenov ideas whereby tangential contact between the plots for heat release and heat loss as a function of temperature is a critical condition beyond which there will be a fundamental change in combustion phenomenology. A summary of such treatments of flashover can be found in Jones, 1993.

A general rule is that flashover occurs when a localised fire attains a heat-release rate of 1 MW. This is only an approximation, however. In the detailed investigation of the fatal Kings Cross fire in London in the late 1980s it was concluded that the heat-release rate at flashover was at least 2 MW, and might conceivably have been as high as 7 MW. In fire protection engineering, 1 MW may be treated as a lower bound below which flashover will not occur. This is illustrated in the example below.

2.1 A retailer applies for permission to install an electrically illuminated advertisement structure, approximating to a cylinder of 2 m diameter, inside a shopping mall. Before such permission can be granted the retailer must ensure that if the structure caught fire it would not cause flashover to the entire mall. If, in the event of fire, it transferred heat solely by radiation from its curved surface at 700 ^{o}C and approximated to a black body, what is the maximum allowable height of the cylindrical structure consistent with the requirement that it would not cause flashover? Use a value of 300 K for the temperature of the surroundings.

Solution

As instructed in the question, we treat the structure as a black body radiating at the combustion temperature T, whereupon:

$$q = \sigma A\,(T^4 - T_o^4) \qquad \text{(W)}$$

where q = rate of heat transfer from structure to surroundings (W), σ = Stefan's constant = 5.7×10^{-8} $Wm^{-2}K^{-4}$, A = area of the radiating surface, T = combustion temperature (K), T_o = surrounding temperature (K).

Flashover requires about 1 MW, therefore:

$$q = 5.7 \times 10^{-8} \times 2\pi \times 1 \times H\,\{973^4 - 300^4\} = 10^6 \quad \text{(W)}$$

where H is the height of the structure $\Rightarrow H = 3$ m

Note

The calculation above takes the surroundings to be at 300 K, room temperature. Of course, as a pre-flashover fire takes its course the surroundings will in fact rise in temperature, not only because of radiation but also because of sensible heat in the post-combustion gas which will be transferred to surrounding surfaces by convection, and this will be considered in the next question.

2.2 The condition for flashover, as already stated, is *tangential* contact between the heat release and heat loss curves, expressible as:

$$R = L \quad \text{and} \quad \mathrm{d}R/\mathrm{d}T = \mathrm{d}L/\mathrm{d}T$$

where R (W) is the heat release rate by the combustion process and L (W) the heat transfer rate to the surroundings. Usually L has a fairly simple functional form such as

$$L = \chi\, A(T - T_O)$$

where χ ($W\,m^{-2}K^{-1}$) is an overall heat transfer coefficient incorporating radiation and convection and A is the area over which heat transfer from pre-flashover fire to surroundings occurs. The functional form of R is more complicated, usually being of the Arrhenius form:

$$R \propto \exp(-E/\mathrm{R}T)$$

where E ($J\,mol^{-1}$) is the activation energy and R the gas constant (8.314 $J\,K^{-1}mol^{-1}$). However, it is sometimes valid across a limited temperature range to substitute T^n for the exponential, whereupon the expression for heat release rate becomes:

$$R = \phi\, T^n$$

where ϕ is a temperature–independent coefficient incorporating the heat of reaction. On the basis of this expression for R and the expression for L previously given, deduce the condition for flashover in terms of T, T_o, χ, A and n.

Solution

$R = L \Rightarrow \phi T^n = \chi A(T - T_o) = 10^6$ (W)

$dR/dT = dL/dT \Rightarrow n\phi T^{n-1} = \chi A$

Dividing one equation by the other and rearranging:

$$(T/n) = T - T_o = 10^6 / \chi A$$

2.3 Return to the cylindrical structure in question 2.1, and consider a radius of 1 m and a height of 3 m. We retain the assumption that heat transfer is purely radiative to the exclusion of convection whereupon χ is a radiation heat transfer coefficient expressed in $Wm^{-2}K^{-1}$. According to Holman (2002),

$$\chi = \sigma (T^2 + T_o^2)(T + T_o)$$

where σ is the Stefan Boltzmann constant.

On this basis calculate the flashover temperature for a surrounding temperature of 300 K, as before.

Solution

Rearranging the condition for flashover:

$\chi A(T - T_o) = \sigma A(T^2 + T_o^2)(T + T_o)(T - T_o) = 10^6$ (W)

$[(T^2 + T_o^2)(T + T_o)(T - T_o)] = 9.3 \times 10^{11}$ (K^4)

$[(T^2 + T_o^2)(T^2 - T_o^2)] = 9.3 \times 10^{11}$ (K^4)

$(T^4 - T_o^4) = 9.3 \times 10^{11}$ (K^4)

$$\Rightarrow T = 980 \text{ K } (707\ ^{\circ}\text{C})$$

Note

That this is almost exactly the same as the value in Exercise 2.1 is an indication of the equivalence of the two approaches once it has been assumed that heat transfer to the surroundings is purely radiative. Note that the value of n required to make the model which was developed in the previous question work is:

$$980/(980 - 300) = 1.44$$

In other applications (see Jones and Wake, 1990) where the exponential is replaced by T^n the value of n is usually somewhat higher than this. Such applications are the calculation of critical conditions for explosion of thermally unstable substances. Consistency between the working above and that in Exercise 2.2 can be confirmed by calculating χ in each case.

From Exercise 2.2:

$$\chi = \{10^6/[(980 - 300) \times 2\pi \times 1 \times 3]\} = 78 \ \text{W m}^{-2}\text{K}^{-1}$$

From Exercise 2.3:

$$\chi = \sigma(980^2 + 300^2)(980 + 300) = 77 \ \text{W m}^{-2}\text{K}^{-1}$$

i.e. the values do not differ significantly.

2.4 In the above calculations on the cylindrical structure, heat transfer to the surroundings was taken to be purely radiative. We now extend it to include convection. Calculate the convection heat transfer coefficient h_c for the cylinder at 980 K, using the approximate correlation

$$h_c = 1.31(T_s - T_o)^{1/3} \quad \text{W m}^{-2}\text{K}^{-1}$$

provided by Holman (2002) where T_s is the cylinder surface temperature and T_o the surrounding air temperature, valid for Rayleigh numbers above 10^9. First confirm that the Rayleigh number for the conditions being examined *is* such that the above correlation applies.

Determine to what extent convection would in fact contribute using the expression:

$$h_{total} = h_c + h_r$$

and explain why the calculation might have to be adjusted if the interior of the shopping mall was heated.

Solution

Repeating the calculation of the radiation heat transfer coefficient:

$$h_r = \sigma(T^2 + T_o^2)(T + T_o)$$
$$= 5.7 \times 10^{-8} (980^2 + 300^2)(980 + 300) = 77 \text{ W m}^{-2}\text{K}^{-1}$$

$$\text{Ra} = (g\,\beta\,\Delta T\,x^3/\nu^2)\,\text{Pr}$$

where g = acceleration due to gravity (9.81 m s^{-2}), β = compressibility factor (K^{-1}), ΔT = surface temperature minus surrounding air temperature (K), x = vertical dimension (m), ν = kinematic viscosity ($m^2 s^{-1}$), Pr = Prandtl number (see Chapter 4), all fluid properties at the film temperature T_f given by:

$$T_f = (T + T_o)/2 = 640 \text{ K}$$

For a gas $\beta = 1/T_f = 1.6 \times 10^{-3}$ K^{-1}, $x = 3$ m and from tables of the properties of gases in Holman (2002):

$$\nu = 5.5 \times 10^{-5} \text{ m}^2 \text{ s}^{-1} \text{ and Pr} = 0.7 \Rightarrow \text{Ra} = 7 \times 10^{10}$$

therefore the correlation above is appropriate, giving:

$$h_c = 1.31 \times 680^{1/3} = 12 \text{ W m}^{-2}\text{K}^{-1}$$

$$h_{total} = h_c + h_r = (77 + 12) = 89 \text{ W m}^{-2}\text{K}^{-1}$$

$$h_c / h_{total} = 12/89 = 0.135$$

so although the convection is small in comparison with the radiation it is not totally insignificant. Use of h_r rather than h_{total} in predictive calculations of flashover will err on the safe side: the body is less likely to ignite if the radiative heat loss to the surroundings is augmented by convective heat loss.

In the above calculations we have used the same value for surrounding temperature for the radiation and convection calculations. Air is however transparent to thermal radiation, so for radiation purposes the surrounding temperature is that of the walls of the mall of which the cylinder has a 'view'. If there is no heating, to assume that the air and wall temperatures are about the same is reasonable. If however there *is* heating (e.g. from steam pipes or other devices in or close to the walls) the walls will be hotter than the air and a different surrounding temperature would be required for the radiation calculation.

Note

h_{total} corresponds to χ in solution to Exercise 2.2. In Exercise 2.3, the assumption is made that $h_r \gg h_c \Rightarrow \chi \approx h_r$ and this approximation has been retrospectively examined in Exercise 2.4.

Returning to the final part of the previous question, the author's intuition is that slightly warmer walls would hardly affect h_r when the body temperature is as high as 980 K, but a thorough and perceptive analysis of the situation requires that this point is raised even though its subsequent dismissal can be justified.

2.5 A pre-flashover fire approximates to a 1 m cube of uniform temperature 700 °C. It transfers heat to the surroundings at 300 K purely by radiation as a grey body with emissivity (ε) 0.4 at a rate q given by

$$q = \varepsilon\sigma A(T^4 - T_o^4) \quad \text{(W)}$$

where A is the radiating area. Determine q and make a judgement as to whether there is a possibility of flashover.

Solution

For a 1 m cube, the area of one face is 1 m^2 and there are 6 such faces giving A = 6 m^2

$$q = 0.4 \times 5.7 \times 10^{-8} \times 6 \times \{973^4 - 300^4\} = 0.12 \text{ MW; too small for flashover.}$$

2.6 Referring to the previous question, how big would the fire need to be, other things being equal, for flashover to develop?

Solution

Set q = 1 MW ⇒ A = 49 m^2, each side 8.2 m^2, therefore the length of one side would be just under 3 m.

Note

The term 'grey body' introduced in Exercise 2.5 means of course a body having a lower emissivity than a black one but having the same emissivity towards all wavelengths of interest.

A helpful comparison with the situation in the first four Exercises in this section is possible. Flashover conditions were:

temperature: 973 K

area: $2\pi \times 1 \text{ m} \times 3 \text{ m} = 19 \text{ m}^2$

emissivity: 1

In Exercise 2.6, the flashover temperature was also 973 K. However, the area was 49 m^2 and the emissivity 0.4. Assigning subscript a to the situation in Exercises 2.1 to 2.4 and subscript b to the situation in Exercises 2.5 and 2.6,

$$A_a / A_b = 19/49 = 0.4$$

$$\varepsilon_b / \varepsilon_a = 0.4/1 = 0.4$$

The ratio of the areas is therefore equal to the inverse ratio of the emissivities.

Flashover features in later parts of the text, notably that on vehicular fires.

Chapter 3:
Post-flashover fires

Introduction

Once a fire has become established its temperature will not be static. However, where flashover alone is of interest, some models make this assumption. There will certainly not be a temperature rise as rapid as that at flashover but there will be a significant rise during burning followed by a drop (sometimes called the decay period) as combustibles become depleted. A schematic of this is provided by Lie (2002) and it is worth noting that the time to flashover calculated there corresponds well with values calculated in the later section of this text.

National standards for fire temperature as a function of duration apply and have been expressed as various analytical forms, including:

$$T - T_o = 345 \log_{10} (8t + 1)$$

an expression attributed to ISO, where T = temperature (°C) at time t minutes and T_o = temperature (°C) at t = 0. One might intuitively expect the fire load to appear in such a correlation. However, ventilation rather than fuel availability controls burning in the post-flashover regime and the total quantity of fuel is in effect incorporated in the duration, i.e. the point at which $T = T_o$ for the second time. However, this point is not calculable from the above equation which is only for the growth period, not the decay period. Nevertheless T_o is the initial temperature, not the temperature immediately after flashover.

3.1 According to the ISO expression, for how long does a fire need to burn to attain a temperature of 1000 °C? Let T_o = 20 °C.

Solution

$980 = 345 \log_{10} (8t + 1) \Rightarrow (8t + 1) = 10^{(980/345)} = 693 \Rightarrow t = 86$ minutes

Note

Table 4.8.2 in Lie (2002) uses $T_o = 20$ °C and gives 1 hour 30 min for the time to reach 978 °C, in reasonable agreement with the above.

> 3.2a According to the ISO correlation previously used, what will be the rate of temperature rise:
>
> (i) 1 minute after initiation,
>
> (ii) 15 minutes after initiation and
>
> (iii) 90 minutes after initiation?
>
> Express answers in °C per minute.

Solution

$T - T_o = 345 \log_{10}(8t + 1) = (345/2.303) \ln(8t + 1) = 150 \ln(8t + 1)$

$dT/dt = 150 \times 8/(8t + 1) = 1200/(8t + 1)$

(i) After 1 minute, $dT/dt = 133$ °C min^{-1}

(ii) After 15 minutes, $dT/dt = 10$ °C min^{-1}

(iii) After 90 minutes, $dT/dt = 1.7$ °C min^{-1}

Note

The temperature trajectory rises rapidly at first then more slowly as shown in the schematic provided by Lie in 2002.

> 3.2b Extend question 3.2 (a) by calculating the initial rate of temperature rise ($t = 0$).

Solution

$dT/dt = 1200/(8t + 1)$

Putting $t = 0$ gives:

$$dT/dt = 1200 \text{ °C min}^{-1}$$

> 3.3a Lie (2002) provides other equations for temperature rise during fires, including:
>
> $$T - T_o = a\,[1 - \exp(-3.79553\sqrt{t})] + b\sqrt{t}$$
>
> where t is time (hours), a = 750 °C and b = 170.41 °C hour$^{-0.5}$. Derive from this the rate of temperature rise at $t = 90$ minutes (1.5 hours).

Solution

$d/dt\,[\exp(-3.79553\sqrt{t})] = -\,(1.90/\sqrt{t})\,\exp(-3.79553\sqrt{t})$

Differentiating term by term the equation in the question:

$dT/dt = 0.5b/\sqrt{t} + (1.9a/\sqrt{t})\,\exp(-3.79553\sqrt{t})$

$= 85.205/\sqrt{t} + (1425/\sqrt{t})\,\exp(-3.79553\sqrt{t})$

Putting $t = (90/60) = 1.5$ hour gives:

$$dT/dt = 81\ ^{o}C\ hour^{-1} = 1.35\ ^{o}C\ min^{-1}$$

which is in good agreement with the value calculated by the ISO correlation.

3.3b Obtain the value of dT/dt at 15 minutes from the equation given in the previous question.

Solution

$dT/dt = 85.205/\sqrt{t} + (1425/\sqrt{t})\,\exp(-3.79553\sqrt{t})$

Putting t = 15 minute = 0.25 hour gives:

$$dT/dt = 598\ ^{o}C\ hour^{-1} = 10\ ^{o}C\ min^{-1}$$

in exact agreement with the prediction of the ISO correlation. The reader can easily confirm that the value at 1 minute calculated from the second correlation that we have considered is 125 $^{o}C\ min^{-1}$, again in quite good agreement with the prediction of the ISO model.

3.4 On the basis of each of the correlations used in this section of the text, calculate the temperature after one hour.

Solution

The correlation:

$T - T_o = a\,[1 - \exp(-3.79553\sqrt{t})] + b\sqrt{t}$

with a = 750 ^{o}C, b = 170.41 ^{o}C and t in *hours*

$$\Rightarrow T = 904^{o}\ C$$

The ISO correlation:

$T - T_o = 345\log_{10}(8t + 1)$ with t in *minutes*

$$\Rightarrow T = 945^{o}\ C$$

Note

The two correlations give good agreement. A thermocouple in a 'tree' arrangement assembled in the fire could probably not distinguish the two.

Chapter 4:
Ignition of solid materials

Introduction

In an accidental fire a solid surface might well receive heat by radiation from something nearby which is flaming. Experimentalists have determined for various widely used materials the critical heat flux (CHF, units kW m^{-2}) required for ignition. Given that the CHF is exceeded, the time to ignition can be calculated. This is illustrated in the calculation below. An article by Tewarson (2002) has been drawn on extensively in what follows.

Tewarson correlates the CHF with the time to ignition of an irradiated surface in the following way.

$$(t_{ig})^{-1/2} = (q' - \text{CHF})/\text{TRP}$$

where t_{ig} = ignition time (s), q' = heat flux (kW m^{-2}) at the receiver and TRP = thermal response parameter (units kW $s^{0.5}$ m^{-2}). The TRP depends *inter alia* on the thermal conductivity and the heat capacity of the material experiencing irradiation. Tewarson also tabulates CHF–TRP pairs for many materials, enabling time to ignition for various irradiances to be given. A number of related examples are given below.

> 4.1 An article of furniture in a room catches fire. The nearest inside wall of the room receives heat by radiation at a rate of 15 kW m^{-2}. The wall is made of a particular timber, the CHF of which is 10 kW m^{-2} and the TRP 140 kW $s^{0.5}$ m^{-2}. How long will it take for the wall surface to ignite?

Solution

$(t_{ig})^{-1/2} = (q' - \text{CHF})/\text{TRP}$

Putting q' = 15 kW m^{-2}, CHF = 10 kW m^{-2}, TRP = 140 kW $s^{0.5}$ m^{-2} gives:

$$t_{ig} = 784 \text{ s (13 minutes)}$$

Note

Flashover is defined as the transition from a localised fire to one involving the whole room. Arguably here we can claim to have calculated a 'time to flashover' in the above example. Once the wall has ignited there will of course be a large increase in the total heat-release rate, another definition of flashover. The calculated time of 13 minutes would probably be long enough for evacuation before flashover though this depends of course on factors including the population of the room and the ease of movement of persons. This theme continues in the next question.

4.2 Imagine that in the previous question the time of 13 minutes to flashover is seen as being too short, perhaps because the room contains elderly people who need some help in evacuating in the event of fire. It is decided that a synthetic material with CHF 13 kW m^{-2} and TRP 200 kW $s^{0.5}$ m^{-2} will be substituted. By how much will this extend the time to flashover if the irradiance is as before?

Solution

$$(t_{ig})^{-1/2} = (q' - \text{CHF})/\text{TRP} \Rightarrow t_{ig} = 10000 \text{ s (2.8 hours)}$$

Note

It is easy to envisage a real circumstance to which the hypothetical situations in the previous two Exercises would conform. A large dwelling house is to be converted to a home for the elderly. There will be at least one room in the building so converted in which the residents are allowed to smoke, and it is estimated that a chair or sofa accidentally ignited by a cigarette will transfer heat to the walls at the stated irradiance value. If the wall of the room is made of timber panels having the CHF and TRP values given in Exercise 4.1, flashover will follow in less than a quarter of an hour. If the timber is replaced by the synthetic material in Exercise 4.2, flashover takes nearly 3 hours and there will of course have been extinguishment measures long before then, which means in fact that flashover will not occur at all. The CHF and TRP values in Exercise 4.2 are intended to correspond to no particular material but are typical values for synthetic materials (Tewarson, 2002).

4.3 Continuing the previous two calculations, it is argued that given the size of the room, the configuration of the furnishings and other such factors including the ventilation rate, the figure of 15 kW m^{-2} for the flux cannot be exceeded. This means that if a material with a higher CHF than 15 kW m^{-2} were used as the wall material, ignition of a piece of furniture could not possibly be followed by flashover. Identify such a material from Tewarson (2002).

Solution

PVC panelling has a CHF of 17 kW m^{-2} (Tewarson 2002)

Note

Though perhaps an unlikely choice of material in the situation being considered, let it be noted that the CHF of Teflon is 38 kW m^{-2}. A material having a low TRP can of course be coated with something else to give it a higher TRP. This point is examined in the next question.

> 4.4 A warehouse is used to store newly printed books. These have a quite significant calorific value (about 15 MJ kg^{-1}) so a warehouse full of them would have a high fire loading. Insurers require that instead of simply being stored in cardboard boxes, which would be the normal procedure, stockpiles of books are covered in corrugated paper which has been coated with a polyester/fibre glass material raising the TRP. The TRP of the paper without coating is 181.5 kW $s^{0.5}$ m^{-2} and with coating (Tewarson, 2002) this rises according to:
>
> $$\text{TRP} = 181.5 + 24.5w$$
>
> where w is the weight of the coating expressed as a percentage of the weight of the corrugated paper. The CHF of the unprotected corrugated paper is 10 kW m^{-2}. It is required that the coated corrugated paper can withstand an irradiance of 25 kW m^{-2} for 15 minutes. What extent of coating with the polyester/fibre glass material would be required?

Solution

It is debatable whether we should use the CHF for the unprotected corrugated paper for the coated paper but clearly to do so will err on the safe side.

Using:

$(t_{ig})^{-1/2} = (q' - \text{CHF})/\text{TRP}$

the TRP required to meet the criterion in the question is calculable.

$\text{TRP} = (q' - \text{CHF})\,(t_{ig})^{1/2} = 450 \text{ kW s}^{0.5} \text{ m}^{-2}$

$$450 = 181.5 + 24.5\,w \Rightarrow w = 11\%$$

Note

The above Exercise considers a composite material, paper and the coating. In the previous Exercises in this section we considered a single material and regarded that as having a thickness-independent value of the TRP. The TRP of a material does in fact depend on the thickness, but above a certain minimum thickness (about 5 mm) the dependence is fairly weak. (The thickness is of course always much smaller than the dimension in either of the other directions.) Nevertheless the calculations in the earlier part of this section would benefit from refinement once a thickness of material had been determined. Refinement is also possible in relation to the CHF of the paper once coated.

> 4.5a The CHF of high-density polyethylene is 15 kW m^{-2} and TRP 325 kW $s^{0.5}$ m^{-2}. If a sheet of this material is irradiated at 35 kW m^{-2} how long will it take for ignition to occur?

Solution

$$(t_{ig})^{-1/2} = (q' - \text{CHF})/\text{TRP} \Rightarrow t_{ig} = 265 \text{ s}$$

> 4.5b A panel is made from the material in the previous question, high-density polyethylene. The density of the material is 950 kg m^{-3}. The panel is 5 cm thick. Calculate the mass of 1 m^2 of the screen material. If it is irradiated at 35 kW m^{-2} for 265 s, how much of the material will be converted to monomer given that the heat required for such conversion (frequently referred to as the heat of gasification) is 2 MJ kg^{-1}?

Solution

$$\text{Weight of 1 m}^2 \text{ of the material} = (950 \times 1^2 \times 0.05) \text{ kg} = 47.5 \text{ kg}$$

$$\text{Mass gasified on irradiation} = [35000 \times 265/(2 \times 10^6)] \text{ kg} = 4.6 \text{ kg}$$

Note

4.6 kg of ethylene is 164 mol or about 4 m^3 at 1 bar, 15 °C. The amount actually in the burning zone at ignition will be much smaller than this. Clearly losses due to ethylene diffusion are occurring. The diffusion coefficient of ethylene in air is of the order of 10^{-5} m^2s^{-1} therefore during the irradiance time ethylene will diffuse a distance of about:

$$(2 \times 10^{-5} \times 265)^{0.5} = 0.07 \text{ m}$$

> 4.5c On the hypothesis that ignition occurs when the proportion of monomer (in this case ethylene) in the gas contacting the polymer surface is in a half-stoichiometric[1] proportion with air, what will be the proportion of ethylene?

[1] It is a widely used rule of thumb, especially in the prediction of flash points, that the lower flammability limit of a hydrocarbon corresponds to half-stoichiometric.

Solution

The equation is:

$$C_2H_4 + 3O_2 + 11.3\ N_2 \rightarrow 2CO_2 + 2H_2O + 11.3\ N_2$$

Proportion of ethylene for ignition = 0.5/(1 + 3 + 11.3) = 0.033

Note

It is helpful to draw together what we have learnt from the three parts of this Exercise. Part (a) was a routine application of CHF and THP, and in part (b) the heat of gasification was introduced. It was shown in part (b) that about a tenth of the polymer will be converted to vapour during the irradiation time and that a great deal of this will diffuse away unreacted. It was shown in part (c) that a proportion of the monomer of 3.3 % in the gas phase above the surface of the polymer being irradiated is necessary for there to be an ignitable mixture.

> 4.6a Return to the question on newsprint in the section on fire loads. Once newsprint ignites, flaming propagation is rapid therefore insurers need to satisfy themselves that if a fire did begin elsewhere in the room there would be sufficient ignition delay for evacuation. If a minimum of 15 minutes is required for evacuation what would be the maximum flux which the newsprint could experience consistently with this requirement? Newsprint has a CHF of 10 kW m^{-2} and a TRP of 108 kW $s^{0.5}$ m^{-2}.

Solution

From $(t_{ig})^{-1/2} = (q' - \text{CHF})/\text{TRP}$

$$q' = 13.5\ \text{kW m}^{-2}$$

> 4.6b With reference to the conditions in the previous question it is reasoned that were a fire to begin elsewhere in the room the flux received by the stock of newsprint would be about a fifth of that emitted at the fire. At what temperature of the initial fire would there be ignition of the newsprint? Approximate the initial fire to a black body.

Solution

Using the CHF:

Flux at the fire = 5 × 13500 W m^{-2} = σT^4 where σ is the Stefan–Boltzmann constant

$$\Rightarrow T = 1043\ \text{K}\ (770\ ^{\circ}\text{C})$$

Note

While there is nothing wrong with expressing the incident flux as a proportion of the emitted flux as in the above Exercise, we have to understand that in the former the area is referred to the receiver and in the latter unit area is referred to the source. Failure to recognise this has sometimes led to concepts of 'conservation of flux' which are of course erroneous. Energy is conserved: fluxes can be added to or subtracted from each other but no conservation law applies.

4.7 The TRP is relevant not only to ignition but also to whether there will be subsequent propagation. The Fire Propagation Index (FPI) is defined as:

$$\text{FPI} = 750\, Q^{1/3}/\text{TRP}$$

where Q is the heat-release rate per unit width (kW m^{-1}), propagation being vertical and upward. In some materials, propagation hardly goes beyond the ignition zone while in others there is accelerating propagation. The latter requires an FPI value of about 20 or higher.

Return to question 4.5a, which considers high-density polyethylene having TRP 325 kW $s^{0.5}$ m^{-2}. What value of Q would be required for rapid propagation?

Solution

Inserting the numbers with FPI = 20,

$$Q = 650 \text{ kW m}^{-1}$$

Note

The equation given in the Exercise above is for a thermally thick sample, that is, one of physical thickness such that when the flame traverses a particular part of the surface, that of the *facing* surface at the same height does not show any thermal response at all. A few centimetres thickness is usually sufficient to ensure this. We continue the theme of thermal thickness below. Unlike the physical thickness, the thermal thickness is not single-valued; it depends on the heat flux being experienced. A closely related quantity is the thermal penetration depth (Babrauskas, 2003) which we consider in the next Exercise. We first need to introduce the quantity thermal diffusivity, usually represented by the symbol α, defined

$$\alpha = k / c\rho \text{ m}^2\text{s}^{-1}$$

where k = thermal conductivity (W $m^{-1}K^{-1}$), c = heat capacity (J $kg^{-1}K^{-1}$) and ρ = density (kg m^{-3}). There are in fact three quantities relevant to transport processes, all of which have units m^2s^{-1}:

- If *heat* is being transported, the quantity is referred to as the thermal diffusivity α
- If *mass* is being transferred, the quantity is referred to as the diffusion coefficient D
- If *momentum* is being transported, the quantity is referred to as the kinematic viscosity ν

The kinematic viscosity ν is defined as:

$$\nu = \mu / \rho \ \ m^2 \ s^{-1}$$

where μ is the dynamic viscosity, measured in kg $m^{-1}s^{-1}$.

Clearly the quotient of any two of these quantities will be dimensionless. Two such dimensionless numbers are in wide use:

$\nu / \alpha = Pr$, the Prandtl number, named after Ludwig Prandtl

and

$\alpha / D = Le$, the Lewis number, named after Bernard Lewis

Historical aside: Ludwig Prandtl died in 1953. Bernard Lewis died as relatively recently as 1993, at a very great age, so many combustion experts still in the job (including the author) can claim to have made his acquaintance. Lewis co-founded the Combustion Institute in 1954, currently an international body of enormous size and influence.

4.8a The point made above is that the thermal thickness depends on the heat flux being experienced, and that a closely related quantity is the thermal penetration depth δ_p. A body is thermally thick if the temperature at the thermal penetration depth does not rise as a result of irradiation of the surface at a specified rate for a specified time.

The thermal diffusivity α of polystyrene is approximately 1×10^{-7} m^2s^{-1}. A slab of polyethylene is irradiated for 5 minutes. What is the thermal penetration depth

(i) if $\delta_p = 4\sqrt{(\alpha t)}$; where t is the irradiation time,

(ii) if $\delta_p = 1.13 \sqrt{(\alpha t)}$?

Thermal penetration depth definitions from Babrauskas (2003).

Solution

$$\text{(i) } \delta_p = 22 \text{ mm}; \quad \text{(ii) } \delta_p = 6 \text{ mm}$$

Note

Criterion (i) defines the penetration depth as being that at which the temperature rise at time t is 0.5 % of the rise at the surface, and is sometimes considered to be too conservative. Let us therefore use 6 mm as the thermal penetration depth. The material is thermally thin if irradiation for five minutes causes a significant temperature rise 6 mm below the surface; otherwise it is thermally thick. This leads to the point made previously that *whether a body is thermally thick or thermally thin depends on the heat flux it is experiencing.*

4.8b Using the mean of the two δ_p values calculated previously, i.e. 14 mm, use the semi-infinite solid model (Holman, 2002) for constant surface flux to estimate at what irradiation rate q (W m^{-2}) the polystyrene slab will cease to be thermally thick.

Solution

The analytical solution to this problem is

$$T(x, t) - T_i = [2q(\alpha t/\pi)^{1/2} / k]\exp(-x^2/4\alpha t) - qx / k\,\{1 - \text{erf}\,[x / 2\sqrt{(\alpha t)}]\}$$

where thermal conductivity $k = 0.12$ W m^{-1}K^{-1}, distance below the surface $x = 14 \times 10^{-3}$ m and the initial temperature T_i is taken to be 25 °C. This is of course, before the application of the flux, the temperature at every point in the slab.

We have all the quantities we need[2] to employ the above equation to determine q provided that we have on the left hand side of the equation an operational definition of thermal thickness in terms of the temperature difference. We can justify any such definition in retrospect by calculating the surface temperature after 5 minutes and obtaining the temperature rise at the target depth from the temperature rise at the surface.

Let us arbitrarily put: T(14 mm, 5 minutes) = 27 °C.

$$2 = 0.0101q - 0.0093q$$

$$q = 2.5 \text{ kW m}^{-2}$$

The surface temperature after 5 minutes ($x = 0$) is calculable as 154 °C.

[2] The error function required ('erf') is provided by Holman (2002) amongst other sources.

The temperature rise is 51 °C and the rise at 14 mm depth is 3 % of this, making our arbitrary criterion reasonable.

4.8c In the irradiation of a polystyrene slab considered in Exercises 4.8a and 4.8b, how long will it take for the surface to reach 100 °C?

Solution

Putting 75 °C for the temperature difference on the left hand side of the governing equation and $x = 0$ since it is the surface temperature which is being calculated,

$$75 = 2q\,(\alpha t / \pi)^{1/2} / k \Rightarrow t = 106 \text{ seconds}$$

4.9 If a layer of newsprint is receiving solar flux of 1000 W m^{-2} what will be its thermal penetration depth after:

(i) 5 minutes (ii) 1 hour

according to the two criteria given in question 4.8a. Use a value of 1.4 $\times 10^{-7}$ $m^2 s^{-1}$ for the thermal diffusivity.

Solution

$$\delta_p = 4\sqrt{(\alpha t)} \Rightarrow \delta_p = 3 \text{ cm after 5 min, 9 cm after 1 hour}$$

$$\delta_p = 1.13\,\sqrt{(\alpha t)} \Rightarrow \delta_p = 0.7 \text{ cm after 5 minutes, 2.5 cm after 1 hour.}$$

4.10 Solar flux, which is of course periodic, has a value of 1400 W m^{-2} at high sun on a clear day. A conservative criterion is that a layer of newsprint will ignite if its surface temperature gets to 170 °C. Is it possible for such a layer initially at 25 °C to be ignited by solar flux? Use a value of 0.24 W $m^{-1}K^{-1}$ for the thermal conductivity.

Solution

Returning to the expression for the temperature in a semi-infinite solid with constant flux:

$$T(x, t) - T_i = [2q\,(\alpha t/\pi)^{1/2} / k]\exp(-x^2/4\alpha t) - qx / k\{1 - \text{erf}\,[x/2\sqrt{(\alpha t)}]\}$$

At the surface (x = 0) this becomes:

$$T(0, t) - T_i = [2q(\alpha t / \pi)^{1/2} / k]$$

If we let $q = 1400$ W m^{-2} for the whole duration that is obviously erring on the safe side and this, using the value of the thermal diffusivity given in the previous question, gives:

t = 3470 s, just under an hour.

Note

Even though we erred on the safe side by using the maximum value of the flux, the indication from the answer is that newsprint can easily be ignited by solar flux, having regard to the fact that daylight is experienced for many more than one hour in twenty-four! Note however that the above treatment has erred on the safe side in two other ways:

(i) By assuming that the newsprint layer absorbs all of the radiation incident upon it, reflecting none. If this were so the layer would of course look black and such 'newsprint' would not be suitable for its intended purpose; indeed, newsprint is bleached in manufacture to make it white. This leads us to the important and frequently misunderstood connection, or rather, lack of any strong connection, between black body behaviour and black colour. If the newsprint looks white (or any colour other than black) it must be reflecting some radiation in the visible wavelength range, but thermal radiation encompasses wavelengths orders of magnitude higher and wavelengths orders of magnitude lower than those in the visible range. It is therefore quite possible for the newsprint to be absorbing nearly all of the energy of the incident thermal radiation notwithstanding its white appearance.

(ii) By ignoring convective losses from the newsprint layer to the air with which it is in contact.

The above calculation is therefore very conservative.

Chapter 5:
Vehicular fires

Introduction

One sees surprisingly few papers in the mainstream fire journals on vehicular fires. A small car such as the classic Mini would, with a full tank, have been carrying about 20 kg of gasoline capable of releasing when burnt about a GJ of heat. Readers will no doubt be familiar with the dismal story of the Ford Pinto[3] in the 1960s, where a design fault that made the car susceptible to fuel ignition in the event of even a fairly mild hit from the rear had been identified before release of the model onto the market, but had not been rectified. At about the time of the Pinto there was worldwide activity in what were then new ideas in car design and configuration, including engines at the rear and sideways orientation of engines at the front. In the development of new designs and configurations the safety of the positioning of the petrol tank within the vehicle structure, in particular its response to collision, is an important factor.

5.1a In a controlled laboratory test (Babrauskas, 2002) a motor car was ignited and allowed to burn out completely, the heat-release rate (HRR) from the burning car being measured continuously. The HRR history can be roughly represented by the schematic diagram below. Calculate the total heat released between ignition and complete burnout.

[3] The Ford Pinto has a place in engineering history bringing with it 'negative prestige'. Further details on http://www.fordpinto.com/blowup.htm

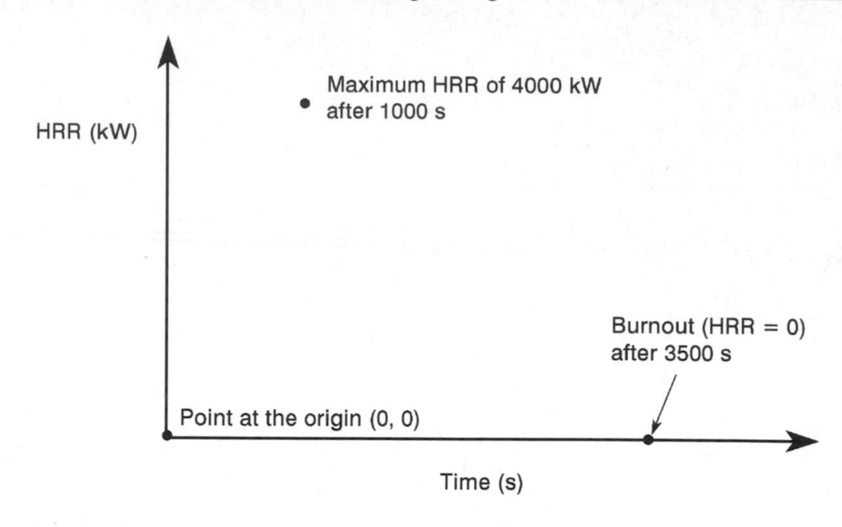

Solution

The points in the schematic graph form a triangle.

Total heat released = area inside the triangle = half base × height

$$= 0.5 \times 4 \times 10^6 \times 3500 = 7 \times 10^9 \text{ J (7 GJ)}$$

5.1b The combustible components of the car in the previous question are chiefly polymer materials in the upholstery, carpeting and interior fittings such as the dashboard. The paint also makes a contribution.

Estimate the weight of such materials before ignition. If the car weighed a tonne before ignition, what proportion of the weight is accounted for by these materials.

Solution

Some polymer materials have of course higher calorific values than others depending on whether the monomer is a hydrocarbon (as with polythene) or a substituted one (as with PVC). Using a value of 35 MJ kg^{-1} to represent the polymers in the vehicle:

$$\text{Weight} = (7 \times 10^9 / 35 \times 10^6) = 200 \text{ kg}$$

$$\text{Proportion of the weight} = 200/1000 = 20\ \%$$

Note

This answer is intuitively sensible. Although the steel from which the body shell is made is of course oxidisable it is doubtful whether temperatures would have been high enough for there to have been a major contribution to the heat release from this. Temperatures of not less than 1500 K would be required for the steel to ignite and pre-flashover fires (see Exercise 5.2b) simply do not approach such temperatures. The reader will have observed that the body shell of a burnt-out car is intact though heavily stained, partly from deposition and partly from surface oxidation.

5.1c If the car in the previous two parts of the question has a boot of capacity 1 m^3 which is filled with unpacked clothing (predominantly cottons) to a bulk density of 50 kg m^{-3}, how much will this add to the combustible component of the car?

Solution

Calorific value of cotton ≈ 16 MJ kg^{-1}; Weight of the clothing = 50 kg

Heat which this quantity can release on burning = 1 GJ (approximately)

Quantity of combustible material expressed in thermal terms is raised by 1/7 ≈15%

5.2a The sorts of heat-release rates shown in the Figure in Exercise 5.1a are obviously sufficient for flashover, but if the car is out-of-doors there is nothing to flash over *to*. If, however, a car is inside a workshop which ignites, flashover is possible. An ignition source can be provided by the hand-held lamp[4] used by a mechanic working on the car. Suppose that a car inside a workshop is so ignited. Using the diagram previously referred to, obtain an approximate estimate of the time to flashover.

Solution

From the figure, 1000 s are required for a HRR of 4 MW to be attained, therefore 250 s are required for 1 MW. If we conservatively assign 1 MW to the heat-release rate at flashover, the time to flashover will therefore be 250 s, just over 4 minutes.

Note

The use of a heat-release rate that increases linearly with time in the calculation is of course an approximation but readers can satisfy themselves

[4] A lamp for such duty will not normally be a fairly harmless torch drawing on batteries but will draw on the mains electricity supply. Ignition from such a source is a recognised hazard amongst professional car repairers.

from the original source (Babrauskas, 2002) that it is not an unreasonable one. Even the best-managed Fire Department could not be relied upon to be present at the fire within 4 minutes of initiation, so the calculation might indicate the need for the workshop to have own extinguishment facility, perhaps a foam. On the other hand a car repair workshop will not be densely populated (and the insurers might well require that only employees enter such an area and that customers use a waiting room) so time of the order of a few minutes could be sufficient for total evacuation of persons. We have to bear in mind that the calculation above is approximate and that a prudent interpretation of the result is: time to flashover would be 'a few minutes'.

The Figure of HRR against time is in fact for a particular car. Some sorts of car in similar tests had higher peak rates of heat release, up to about 9 MW. The time taken for the peak to be achieved was about the same as that for the car which displayed a peak of 4 MW.

5.2b In some treatments of flashover, heat transfer is taken to be just radiative. The convective part is assumed to be relatively small, and pre-flashover fires are typically at around 750 K. If these facts apply to the situation in the previous question and if the flame is approximated to a sphere, obtain a value for the sphere radius immediately before flashover.

Solution

Treating the fire as a black body:

$$\text{Area of the fire at flashover} = 10^6 / (5.7 \times 10^{-8} \times 750^4) = 55 \text{ m}^2$$

$$\text{For spherical geometry, radius} = (55 / 4\pi)^{0.5} = 2.1 \text{ m}$$

Note

Perhaps the significance of the above calculation is its reassurance: the radius calculated is of the order of the dimension of a car.

5.3a Gasolines have an upper flammability limit (UFL) of about 7 % in air at a total pressure of 1 bar (Babrauskas, 2003). What value of vapour pressure of gasoline is required in the tank space above the liquid for the mixture of air and vapour to be below the UFL? The 'Reid Vapour Pressure' of gasolines, always measured at 38 °C, is usually in the range 30–70 kPa. Compare your answer with the pressure range and comment.

Solution

The vapour pressure would need to be:

$$0.07 \times 10^5 \text{ Pa} = 7 \text{ kPa or lower}$$

Note

Some background on the RVP is first required. As explained by the author (Jones, 2003a), the RVP is not *the* vapour pressure for a gasoline as, for a particular temperature, it depends on the space into which the vapour expands. However, for many practical purposes the RVP is considered to be approximately constant. Petroleum companies adjust the RVP seasonally, and regional variations exist depending on climate. A winter value in some parts of the world can be as high as 88 kPa (Babrauskas, 2002).

Could the actual vapour pressure drop to below a quarter of the lowest RVP usually used, which according to our calculation above is the requirement for a hazard? The answer is yes, in principle. Gasoline consists of a mixture of numerous hydrocarbons with a wide boiling range. It is possible for the low boilers to have been significantly evaporated leaving a predominance of high boilers. These will give an actual vapour pressure well below the RVP. Relevant factors include how long the gasoline has been in the tank, the degree of emptying of the tank and temperature swings of the tank in vehicle operation.

5.3b Methanol is an increasingly important gasoline substitute. Its upper flammability limit is 36.5 % and its lower flammability limit 6.7 %. Its vapour pressure at 38 °C is 30.9 kPa. Repeat the above calculation for a vehicle using neat methanol as fuel

(i) with the tank at 38 °C and

(ii) with the tank at 5 °C.

It will be necessary to consult a source such as Perry's Chemical Engineers' Handbook to obtain a value for the vapour pressure at 5 °C.

Solution

At 38 °C the proportion is 30.9 %, below the UFL so the gas/vapour mixture in the space above the liquid surface would be explosive. If the tank contents temperature were 5 °C the vapour pressure of the methanol would be 5 kPa, giving a proportion of 5 % which is below the lower flammability limit. The gas/vapour in the space above the liquid would not be in the explosive range.

Note

With gasolines it is only a matter of whether the mixture can approach the flammability range from the high side. One would not expect the gas/vapour in the tank to be below the *lower* flammability limit. With methanol the gas/vapour composition can approach the explosive range from the high side or from the low side. With pure methanol there is no control of the vapour pressure for climatic conditions as there is with gasolines and this is a clear safety disadvantage (as well as an operational one). The temperature range 5–38 °C can easily be exceeded depending on whether the vehicle is, for example, in Canada or in Mexico!

5.4a When liquid fuel leaks and the surface onto which it leaks does not confine it, the area of the resulting patch of liquid can be estimated from empirical equations one of which (Gottuck and White, 2002) is:

$A\ /\ V = 14$ m^2 litre^{-1} for a spill of less than 95 litre (25 US gallons).

If the liquid is gasoline it is well above its fire point at ordinary temperatures and will respond to a heat source such as a spark by igniting, flaming propagation ensuing.

A passenger motor car involved in a road accident loses its entire gasoline holding of 6 US gallons and this spreads naturally without there being containment by kerbs or other structures. What area will the resulting patch of liquid be? If the accident is at a time of day such that solar flux is 1000 W m^{-2} how long will it take for all of the leaked gasoline to evaporate? Use the following data:

density of the leaked liquid = 750 kg m^{-3}

heat of vaporisation = 360 kJ kg^{-1}

Solution

6 US gallons = 23 litre approx

Area = $14 \times 23 = 320$ m^2

Mass of liquid leaked = $0.023 \times 750 = 17$ kg

Heat required to evaporate this = $360 \times 17 \times 10^3 = 6 \times 10^6$ J

Rate of heat supply from the sun = $1000 \times 320 = 3.2 \times 10^5$ W

Time required for total evaporation = $(6 \times 10^6/3.2 \times 10^5) = 19$ s

Note

The time taken for solar flux to evaporate the spill is therefore very short and provided that no ignition source took effect during that time there would be no fire due to the leaked gasoline. The heat of vaporisation used is actually that of pure octane. The patch would not have a uniform initial thickness: at the thickest part the value is likely to be of the order of a millimetre.

The value of 1000 W m^{-2} is also high, signifying a time of day close to high sun and also minimum attenuation due to the weather conditions.

5.4b Suppose the accident in the previous part of the question occurs at night when there is no solar flux. The spilt liquid once dispersed is at a temperature of 5 K above the surrounding air from which it receives heat by natural convection. How long will it take the gasoline to evaporate under these conditions? Use the expression:

$$h = 1.4\,(\Delta T / L)^{1/4}$$

where h = convection coefficient (W $m^{-2}K^{-1}$), ΔT = temperature difference between the air and the horizontal pool surface and L is the horizontal dimension (which it will be necessary to estimate).

Solution

Area = 320 m^2 as before, so a reasonable estimate of L is $\sqrt{320}$ m = 18 m

$$\Rightarrow h = 1.0 \text{ W m}^{-2}\text{ K}^{-1}$$

Rate of heat supply = 1 × 320 × 5 = 1600 W

Time require for evaporation = (6×10^6 / 1600) = 3750 s (approx 1 hour)

5.4c Repeat Exercise 5.4b on the basis that the patch of spilt gasoline approximates more closely to a circle than to a square.

Solution

In the event the horizontal dimension is the diameter D:

320 m^2 = $\pi D^2 / 4 \Rightarrow D = 20$ m

$$h = 1.4 \times (5/20)^{1/4} = 0.99 \text{ W m}^{-2}\text{ K}^{-1}$$

i.e. No difference at all.

Note

Evaporation takes much longer at night. If there is a wind, the natural convection considered in parts (b) and (c) might be augmented by forced convection, giving a total convection coefficient about half an order of magnitude higher than that for natural convection only and reducing the evaporation time accordingly.

5.5 One reason amongst others why the structural soundness of a vehicle is important to its roadworthiness is that in the event of a collision, the metal panels will absorb the energy and protect the fuel tank from rupture.

Suppose that a vehicle with major corrosion is nevertheless in use and experiences a hit which causes fuel tank contents to leak catastrophically, that is, all at once. If there is ignition what will be the diameter of the resulting fireball:

(i) if the tank was holding 7 (Imperial) gallons or

(ii) if the tank was holding 2 gallons?

Use a value of 750 kg m^{-3} for the density of the gasoline and the following correlation for fireball diameter:

$$D = 5.8\ W^{1/3}$$

where W is the mass of leaked fuel in kg and D the fireball diameter in m.

Solution

$$7 \text{ Imperial gallons} \equiv 0.0318 \text{ m}^3 \text{ weighing } 24 \text{ kg} \Rightarrow D = 17 \text{ m}$$

$$2 \text{ gallons weigh } 6.9 \text{ kg} \Rightarrow D = 11 \text{ m}$$

Note

Either fireball would engulf the car. Perhaps the above calculations would serve as an object lesson in the cause of elimination of shonky roadworthiness certificates!

The correlation for the fireball diameter, taken from the SFPE Handbook, is one of a number of such correlations. The reader might like to track down others and compare results. Results from different correlations will usually be within about ± 20 %.

Chapter 6: Overpressures

6.1 Propane ignites in an unvented space, the total pressure being initially 1 bar. The initial proportions of air and propane are stoichiometric, and combustion occurs according to:

$$C_3H_8 + 5\ O_2 + 18.8\ N_2 \rightarrow 3\ CO_2 + 4\ H_2O\ \text{(vapour)} + 18.8\ N_2$$

Estimate the final pressure in the limit where the space is perfectly insulated, therefore the isochoric adiabatic flame temperature of 2600 K applies.

Solution

Applying the ideal gas law to the initial (subscript 1) and final (subscript 2) states:

$$P_1V_1 / n_1T_1 = P_2V_2 / n_2T_2 = \mathrm{R}$$

with conventional symbols. Rearranging and noting that 'non-vented' means $V_1 = V_2$, and letting the initial temperature be 298 K:

$$P_2 / P_1 = (n_2 / n_1)\ (T_2 / T_1)$$

$$P_2 / P_1 = 25.8 / 24.8\ (2600 / 298) = 9.1\ \text{bar}$$

Note

Let us examine the conditions in the question. The adiabatic approximation might be justified if the walls of the space containing the explosive gas mixture have a high thermal inertia as indeed they are likely to. An actual explosion pressure of 9 bar is equivalent to an overpressure of 8 bar and this is much higher than would ever be encountered in hydrocarbon accidents. The above question therefore relates to something of the nature of a laboratory test in a steel vessel. The above example considers an enclosure occupied by flammable gas so that the entire space contains a mixture within the flammable range. This will seldom be so in room fires and explosions. A localised volume of leaked flammable gas will ignite and the pressure of the resulting deflagration will not be greatly in excess of atmospheric. It will propagate at a speed of up to about 1 m s^{-1} until it encounters a surface and the damage done, if any, will depend upon how long the pressure remains in excess of atmospheric.

In fact an overpressure of just 0.1 bar can cause structural damage to a building and one of 0.01 bar can cause glass damage. The following example illustrates this.

6.2 A glass window, sides of length 0.5 m, is subjected to an overpressure of 0.01 bar. Determine the weight that would need to be placed on the glass if horizontally orientated to produce an equivalent effect.

Solution

0.01 bar = 10^3 N m^{-2}, giving a total force of $10^3 \times 0.5^2$ Newton or 250 Newton, equivalent to: (250/9.81) kg = 25.5 kg

Note

This approach though basically valid oversimplifies the effects of overpressure because there can be dynamic effects during the rise to maximum overpressure.

The author has discussed overpressure elsewhere (Jones, 1993) so this section will be restricted to these two Exercises. The overpressures theme is however continued in a subsequent section on detonations.

Chapter 7:
Completeness of combustion

Introduction

For the generalised reaction,

fuel + oxygen → combustion products.

The actual heat release expressed as an enthalpy change is known for very many fuels which are single chemical compounds. For example, it is 889 kJ mol^{-1} (≡ 55.6 MJ kg^{-1}) for methane. For fuels which are not simple compounds, e.g. a petroleum fraction or a piece of wood, reliable values of the heat release can be obtained by calorimetry. Such simple calorimetric techniques may be either constant volume, yielding the internal energy change (ΔU) as the heat of reaction, or constant pressure, yielding the enthalpy change (ΔH). The difference is seldom significant and of no real interest to the fire protection engineer. What is important in such work is that there shall be complete combustion otherwise the apparent heat value will not mean anything, being dependent on an unknown extent of reaction. That is why, for example, in a bomb calorimeter brute force is applied by having an atmosphere of pure oxygen — not air — of several bar.

In accidental fires (and indeed in most 'friendly' combustion applications) the oxidant is air at 1 bar and complete combustion cannot be assumed. Incomplete combustion will result in one or all of the following:

Smoke

Carbon monoxide

Unburnt fuel

A number of numerical examples will bring out these ideas.

7.1 Refer to question 5.4a in the Chapter on Vehicular fires. 17 kg of gasoline had leaked as a result of fuel tank rupture. If this ignites, calculate the mass of smoke:

(i) If the gasoline is approximated to n-octane, smoke yield factor 0.038 g g^{-1} (Tewarson, 2002)

(ii) If the gasoline is approximated to benzene, smoke yield factor 0.181 g g^{-1} (Tewarson, 2002)

Solution

(i) With n-octane as the model compound, smoke yield = 0.038×17 kg = 650 g

(ii) With benzene as the model compound, smoke yield = 0.181×17 kg = 3 kg

Note

The smoke yield factor is not a hard number, having a significant dependence on ventilation. The values in Tewarson (2002) are unified values on the basis of an ASTM standard for comparative purposes. Note also that benzene gives a much higher smoke yield. The petrol having leaked will be neither pure octane nor pure benzene but it might have a preponderance of alkanes or of aromatics; this will depend on the nature of the crude. A gasoline originating from an aromatic-base crude will, if burnt under non-premixed conditions, be more productive of smoke than a gasoline derived from an alkane-base crude.

Smoke in the above example should be taken to mean only the particulate matter. There is another, more widely applied, definition of smoke whereby it means the particles plus the suspending gas which might well contain, in addition to carbon dioxide, nitrogen and (possibly) excess oxygen, some unburnt or pyrolysed fuel.

Finally we make some comments on the term 'calorimetry'. Simple calorimetry to measure heats of combustion is straightforward both to perform and to interpret. Nowadays there are many very advanced calorimeters that measure heat-release rates. An example is the cone calorimeter, developed by Dr V. Babrauskas. This has been the basis of a huge amount of useful information on fire hazards from such things as furnishings and clothing. Sometimes calorimeters are developed at particular centres for fire research and take the name of that centre: an example is the OSU[5] apparatus.

Results for such devices require precise knowledge of combustion conditions within them and application of such knowledge in interpreting the results. In other words, there is an apparatus dependence of the raw data

[5] Ohio State University.

and such data can only be translated into fundamental information if the effects of the experimental conditions are accounted for. An example is the point that in a cone calorimeter the radiative heat transfer is much larger than the convective though the latter is not necessarily negligible; see Jones, 2003b. These ideas will be developed in the Exercises that follow.

7.2 Ethane, the second compound in the alkane series (formerly known as the paraffin series), is a widely used fuel gas. It has a heat of combustion, measurable by simple calorimetry or calculable from thermochemistry, of 47.1 MJ kg^{-1} (ΔH_T). In an advanced calorimeter of the genre described above, the heat measured is the chemical heat of combustion (ΔH_{ch}) which, if the combustion is in any way incomplete, will be less than the value given earlier in the question which is for complete combustion. The chemical heat of combustion is often determined indirectly from oxygen consumption using a paramagnetic analyser.

In such a test with ethane as fuel (Tewarson, 2002) the chemical heat of combustion is found to be 45.7 MJ kg^{-1}. Now the heat transferred by radiation and that transferred by convection per unit amount of fuel burnt, symbols respectively ΔH_{rad} and ΔH_{con}, are usually, when summed, equal to the chemical heat of combustion, there being no significant conductive heat transfer. This gives:

$$\Delta H_{ch} = \Delta H_{rad} + \Delta H_{con}$$

and:

$$\Delta H_T = \Delta H_{ch} + \text{losses}$$

In the ethane calorimetry the following data apply:

$$\Delta H_{con} = 34.1 \text{ MJ kg}^{-1}$$

$$\Delta H_{rad} = 11.6 \text{ MJ kg}^{-1}$$

$$\text{CO yield} = 0.001 \text{ g g}^{-1}$$

$$\text{Smoke yield} = 0.013 \text{ g g}^{-1}$$

Calculate the percentage of totally unburnt ethane in the test.

Solution

$\Delta H_{ch} = \Delta H_{rad} + \Delta H_{con} = 45.7$ MJ kg^{-1}

$\Delta H_T = \Delta H_{ch} + \text{losses} \Rightarrow \text{losses} = 1.4$ MJ kg^{-1}

Using a value of 282 kJ mol^{-1} (≡ 10.1 MJ kg^{-1}) for the calorific value of CO:

Loss due to the CO per kg ethane burnt = 0.001 kg × 10.1 MJ kg^{-1} = 0.01 MJ

The smoke, being carbon particles, will have a calorific value of about 25 MJ kg^{-1} whereupon:

Loss due to the smoke per kg ethane burnt = 0.013 kg × 25 MJ kg^{-1} = 0.33 MJ

Losses due to CO and smoke jointly = 0.34 MJ

Total loss = 1.4 MJ therefore, working to one significant figure only:

Loss due to unburnt fuel = 1 MJ

Quantity of ethane which could have released this heat = (1/47.1) kg

= 0.02 kg representing 2 % unburnt fuel.

Note

A reader is reminded that calorimeters such as that in the question measure heat-release rates, sometimes for materials for which hardly any chemical analysis information is available. Application to a simple material such as ethane gives useful performance indicators of such a device. A well-characterised solid material might also be used; the example which follows is concerned with such an application.

7.3 Polystyrene is burnt in the calorimeter which features in the previous question and, retaining the notation, results are as follows:

$$\Delta H_{rad} = 15.6 \text{ MJ kg}^{-1}$$

$$\Delta H_{con} = 14 \text{ MJ kg}^{-1}$$

$$\Delta H_{T} = 39.2 \text{ MJ kg}^{-1}$$

$$\Delta H_{ch} = 29.6 \text{ MJ kg}^{-1}$$

Smoke yield factor = 0.135 g g^{-1} CO yield factor = 0.050 g g^{-1}

Determine the quantity of unreacted fuel in the test

Solution

Losses = (39.2 – 29.6) = 9.6 MJ kg^{-1}

Losses due to CO = (0.05 × 10.1) = 0.51 MJ

Losses due to smoke = (0.135 × 25) J = 3.38 MJ

Losses due to CO and smoke = 3.9 MJ

Losses due to unburnt fuel = (9.6 – 3.9) = 5.7 MJ which could be released by (5.7/39.2) = 0.15 kg.

i.e. 15% of the polystyrene completely unreacted.

Note

Such a device might be used to measure heat release rates from an article of clothing or a piece of furniture in which case significantly incomplete combustion is expected. That does not detract from the value of the heat-release measurements made but knowledge of the extent of reaction is useful in planning such tests. The heat-release rate so measured will of course change as the experiment takes its course. The term 'completely unreacted' in the above requires some qualification. There might have been some pyrolysis to hydrogen and simple hydrocarbons such as methane, which are not accounted for in the energy balance as there is no pertinent information in the experimental results. In that event the solid residue remaining would be a char-like mass without the properties of the original polymer.

Note also that with polystyrene, a solid material, ΔH_{rad} and ΔH_{con} are equivalent whereas with ethane ΔH_{rad} is only about one-third of ΔH_{con}. This is the result *inter alia* of the low emissivity of burnt gases compared to those of solid surfaces.

Chapter 8:
Cooling and extinguishment

Introduction

In the first part of this section we shall consider pumping of water, water being the most common extinguishing agent in fire protection. Consider first the centrifugal pump (Geankoplis, 1993; Crowe *et al.* 2001; Rogers and Mayhew, 1980), widely used in fire applications, consisting of an impeller within a casing. The impeller speed may or may not be variable for a particular pump, a typical operating value being 1750 rpm. Kinetic energy imparted to the water by the impeller raises the water pressure, the value at pump outlet referred to as the head pressure. The extent to which this exceeds atmospheric pressure determines the flow rate of water through the pump. Consequently, for any one pump design there is a characteristic curve of developed head against discharge rate. In the selection of a pump for a particular purpose, the impeller diameter is also important. Simple correlations, sufficiently accurate for most engineering purposes, link the quantities characterising a centrifugal pump. The questions which follow utilise some of these.

8.1 A centrifugal pump to be used for fire extinguishment purposes has a fixed rotation speed of 1700 rpm and, at a particular operating head pressure, a delivery of water of 4 m^3 min^{-1}. It is intended to raise the delivery by 15 % to 4.6 m^3 min^{-1}. This has to be achieved by modification to the rotation speed before the pump is put into service. To what value must the impeller rotation speed be raised? By what factor will this raise the head pressure?

Solution

$$\text{delivery rate} \propto \text{rotation speed}$$

Hence the rotation rate must be increased to $(4.6/4) \times 1700 = 1955$ rpm

Head pressure $\propto$ (delivery rate)$^{2.0}$, hence the pressure is raised by the factor:

$$(4.6/4)^2 = 1.32$$

8.2 Moderate increases in the delivery rate can be achieved by adjustment of impeller rate and head pressure as seen above. However, if a change of say half an order of magnitude is required it will be necessary to employ a centrifugal pump with larger impeller blades. So if, as an example, a manifold of pipes supplying water is increased in extent necessitating a much greater delivery rate when the water supply is in operation a pump with larger impeller blades will be required. Calculate the percentage change in size of impeller blades required to produce an increase of a factor of five in the delivery rate.

Solution

$$\text{delivery rate} \propto \text{blade diameter}^{3.0}$$

$\Rightarrow$ a 71% increase in blade diameter for a half-order of magnitude rise in flow rate.

If therefore the blades in the existing pump were 6-inch diameter it would be necessary to consider blades of about 10.5 inch for the replacement pump.

Note

The above calculations appear very simple, but results of such calculations have to be treated with some caution for the following reason. The assumption 'other things being equal' is sometimes implicit and the extent to which it is true in a particular application not always easy to ascertain. Some coverage of the topic uses the term 'homologous pumps', which means that in going from one pump to another characteristics other than the one under consideration remain constant. Our example immediately above is therefore for 'homologous pumps'.

8.3a The centrifugal pump can of course be analysed according to the steady flow equation, there being pressure effects, kinetic energy, potential energy and heat losses to be considered. A first estimate of the power required by a centrifugal pump can be obtained if one considers pressure effects only and neglects the others.

Return to Exercise 8.1 in which 4 m^3 $minute^{-1}$ of water are delivered and suppose that the head pressure is 30 feet of water. Calculate the power developed by the pump. Express your answer both in kW and in horse power.

Solution

If only pressure effects are important and heat developed is neglected, the steady flow equation simplifies to:

$$W = m_f \Delta P / \rho$$

where W is the power (watt), m_f is the fluid flow rate (kg s^{-1}), ΔP is the head pressure (Pa) and ρ the fluid density (kg m^{-3}).

$$4 \text{ m}^3 \text{ min}^{-1} \text{ of water} \equiv 67 \text{ kg s}^{-1}$$

30 feet of water ≡ $(30 \times 12 \times 0.0254 \times 9.81 \times 1000)$ Pa = 89702 Pa

$$W = 6.0 \text{ kW}$$

From Geankoplis (1993), 1 kW = 1.34 hp ⇒ power developed = 8 hp

> 8.3b If in the example above the pump blades are 6 inch in diameter, confirm that the potential energy effects are small in comparison with the pressure effects.

Solution

If the blades are 6 inches in diameter this is the height through which the liquid is raised.

Contribution of this to the power =

$$67 \times 9.81 \times (6 \times 0.0254) = 0.1 \text{ kW}$$

or about 10% of the pressure effect.

Note

The potential energy effect is therefore small but not quite negligible. The kinetic energy contribution is likely to be about the same as the potential. The interested reader can easily confirm this by assuming the water enters the blades at a negligible speed and exits them at about 1 m s^{-1}.

The questions so far in this section have been intended to familiarise readers with the principles of a centrifugal pump when delivering water for fire control, and to provide a feel for related quantities. Readers should be aware that centrifugal pumps also find extensive application to non-aqueous liquids in the process industries. Such pumps are in continuous use so the economics of operation, directly related to the hp, are perhaps more important than for a fire protection application where use is intermittent.

8.4 Consider again the situation in which the water supply is 4 m^3 min^{-1}. Deduce the extinguishment potential i.e. the rate at which it would take up heat if directed at a fire. Assume the liquid to be initially at 30 °C and, once evaporated, to remain at 100 °C.

Solution

The supply rate in weight terms is 67 kg s^{-1} as we saw previously. Now from tables:

Specific enthalpy of the liquid water at 30 °C = 125.7 kJ kg^{-1}

Specific enthalpy of the vapour at 100 °C = 2675.8 kJ kg^{-1}

Heat absorbed in going from liquid at 30 °C to vapour at 100 °C = 2550.1 kJ kg^{-1}

For a supply rate of 67 kg s^{-1}, rate of heat uptake = 170 MW

Note

From a calculation of the same type as Exercise 2.5 on Flashover, a 20 m cube radiating at about 1050 K from all its surfaces would be transferring heat at about that rate so the 4 m^3 min^{-1} of water is equal and opposite to that.

Note that the use of steam tables is more direct and, because it requires no assumption of a single-valued specific heat of water, is more accurate than a calculation which treats the liquid temperature rise and the evaporation separately.

The water vapour will obviously not remain at 100 °C after its extinguishment function is performed but will equilibrate thermally with the burnt gases on mixing with them. This has no bearing on the calculation.

8.5 The previous Exercises in this Chapter have been concerned with the supply of a single stream of water for extinction of a fire already taking place. However, centrifugal pumps are also used to supply water to be reticulated by a sprinkler system. Performance criteria, including water supply per minute per unit area protected by the sprinkler system, are provided by R P Fleming in successive editions of the SFPE Handbook. Consider a sprinkler system with a total rate of supply of 4 m^3 min^{-1} as in previous calculations. Determine the total extent of the floor area that could be protected if the required sprinkler performance is 0.2 g water per minute per ft^2 of floor area.

Solution

The water supply rate translates to 67 kg s^{-1} as we have already seen, which is equivalent to 4020 kg per minute. So for 0.2 g per minute per ft^2 the area is:

$$(4020 \times 10^3/0.2)\ ft^2 = 2 \times 10^7\ ft^2 \equiv 1.9 \times 10^6\ m^2$$

Note

In fine-tuning of sprinkler performance calculations the required supply per unit area decreases with the actual area. Our calculated result above would be acceptable but may be over-cautious. A high fire load due to the presence of strongly combustible materials e.g. organic solvents may increase the required supply.

Return to the storage space 15 m × 15 m in Chapter 1, which converts to approximately 2500 ft^2. According to Fleming's treatment of the subject, the minimum sufficient rate of supply for this is 0.15 g water per minute per square foot.

8.6 Water for fire extinguishment purposes is required to exit a nozzle at a speed of 25 m s^{-1}, the entire flow arrangement being horizontal. The wide end of the nozzle is of 10 cm diameter and the water enters it at 5 m s^{-1}. What must the diameter of the exit nozzle measure? Calculate also the mass flow rate and the pressure drop, using a value of 0.23 for the kinetic energy correction factor.

Solution

The relevant equation, derived from the steady-flow equation (e.g. Cengel and Turner, 2001), is:

$$u_1^2\ (1 - K_L) / 2g + P_1 / \rho\ g = u_2^2 / 2g + P_2 / \rho\ g$$

where u denotes velocity (m s^{-1}), P denotes pressure, K_L is the kinetic energy correction factor, the role of which is to account for frictional losses, and g is the acceleration due to gravity. Subscripts 1 and 2 denote nozzle entry and exit respectively. By continuity, having regard to the fact that the flow is incompressible:

$$u_1 \pi\ (d_1 / 2)^2 = u_2 \pi\ (d_2 / 2)^2$$

$$\Rightarrow d_2 = d_1 \sqrt{(u_1 / u_2)} = 10 \sqrt{0.2} = 4.5\ \text{cm}$$

Mass flow m_f rate calculated from:

$$m_f = 5 \times 1000 \times \pi \times 0.05^2 = 40\ \text{kg s}^{-1}$$

Check that m_f is the same for both sets of conditions, as it has to be to satisfy the continuity condition:

$$m_f = 25 \times 1000 \times \pi \times 0.0225^2 = 40 \text{ kg s}^{-1}$$

$$K_L = 0.23 \Rightarrow \text{head loss } h_L = u_1^2 \ K_L / 2g = 0.29 \text{ m}$$

Returning to the original equation:

$$u_1^2 / 2g + P_1/\rho g = u_2^2 / 2g + P_2 / \rho g + h_L$$

$$P_1 - P_2 = \rho [(u_2^2 / 2) - (u_1^2 / 2)] + \rho g h_L = 302.8 \text{ kPa}$$

8.7 A simpler and less expensive means of sprinkler operation than a centrifugal pump is a diffuser, which of course, is simply a nozzle (see previous Exercise) in reverse.

Water supplied at 1 m s^{-1} has to be decelerated to 2.4 cm s^{-1} for a sprinkler application. If it is received into a horizontal diffuser of 5 cm diameter, to what diameter must the diffuser extend in order to achieve the required delivery speed.

Solution

The statement that the diffuser is horizontal means that potential energy effects are nil. In orientations other than horizontal, this is not necessarily true for liquid flow.

Letting subscript 1 denote entry and subscript 2 denote exit, continuity gives:

$$u_1 A_1 = u_2 A_2$$

since the flow is incompressible.

$$A_2 = A_1 (u_1 / u_2)$$

$$d_2 = d_1 \sqrt{(u_1 / u_2)} \Rightarrow d_2 = 33 \text{ cm}$$

In the following part of this Chapter we look at foam extinguishing agents. A widely used class of foaming agents is protein foam, prepared by blending a liquid protein concentrate with water. The protein constituent is fibre protein similar to the keratin in wool and sometimes obtained from feathers. A protein foam might also contain a fluorinated agent to improve spreading and penetration. There are also aqueous film-forming foams[6] (AFFF) made from water, a foaming compound and a surfactant. See Scheffey (2002) for more details of the composition of such foams. There

[6] Such foams were used at the Buncefield fuel depot fire, Hertfordshire, UK, in December 2005.

are many performance criteria and these are based on tests with pool fires from various liquid fuels. One simple performance criterion is the 'extinguishment application density', that is, the amount of foam required per unit area of the pool for extinguishment, units litre per m^2, called the extinguishment application density. The value of this for a particular fire will depend on whether the nozzle releasing the foam is fixed or not. If the direction can be controlled by a trained operator, the foam will be more effective than if the nozzle is fixed in which case full advantage of the foam will not be obtained. A related example follows.

8.8 A particular foam when applied to a motor gasoline pool fire has an extinguishment application density of 1.34 litre m^{-2} when used with a moveable nozzle (Scheffey, 2002). Refer to Exercise 5.4a in the Chapter on Vehicular fires where a gasoline spill of area 320 m^2 was considered. What quantity of the foam having extinguishment application density 1.34 litre m^{-2} would be required to extinguish a fire resulting from this quantity of gasoline?

Solution

Quantity required = (1.34×320) litre = 430 litre

Note

The extinguishment application density with a fixed nozzle would have been about half an order of magnitude higher.

8.9a Carbon dioxide is an alternative extinguishing agent to water, having the advantage that it can be used for electrical fires and also that, whenever it is used in preference to water or an aqueous foam, there is less non-thermal damage resulting.

Proprietary information on certain CO_2 extinguishers can be found at http://www.dgfire.co.uk/co2.html. One such extinguisher contains 5 kg of CO_2 and discharges over 15.4 seconds. Using the concept introduced previously in this Chapter, that an extinguisher and a fire can be seen as being equal and opposite, how large a fire at 750 °C could such a device reduce to 250 °C and thereby extinguish? Use a value of 1000 J kg^{-1} $°C^{-1}$ for the specific heat of CO_2.

Solution

5 kg of $CO_2 \times (750 - 250)$ °C $\times$ 1000 J kg^{-1} $°C^{-1}$ / (15.4 s) = 0.16 MW

This is of course a fire well into the pre-flashover regime, but one would only use a fixed-charge extinguisher of the type being considered for such so this must not be seen as a limitation. Reticulated carbon dioxide can be used as fire protection where there is the potential for fires releasing heat in the megawatt range.

Note

The equal and opposite concept is an index of the upper limit of extinguisher performance. Note that 0.2 MW is the order of magnitude of heat-release rates in incipient vehicular fires according to the plot previously presented, so for a vehicle to carry one or more such extinguishers is, on engineering principles, sound.

8.9b Consider a layer of material of area 1 m^2 which has ignited. At what temperature will it be radiating heat at a rate equal and opposite to the capability of the extinguisher considered in Exercise 8.9a? Treat the layer as a black body.

Solution

$$1.6 \times 10^5 = 5.7 \times 10^{-8}\ T^4 \Rightarrow T = 1294\ \text{K}$$

8.10a Post-combustion gas is a suitable cooling agent for hot surfaces with the potential to ignite in that it is inert, having been depleted of oxygen and comprising a mixture of nitrogen and carbon dioxide (unless, of course, the combustion used an unusually high degree of excess air). However, such gas has to be cooled itself before it can serve as a cooling agent and this might be process-integrated, for example to heat water.

Post-combustion gas is released at 1100 °C and required to cool to 50 °C. In doing so it is used to heat water from 20 to 85 °C in a heat exchanger at a rate of 0.2 kg s^{-1}. At what rate must the post-combustion gas be supplied if the heat exchange is 80 % efficient and the specific heat of the gas is 1100 J $kg^{-1}K^{-1}$?

The gas, once cooled to 50 °C, is required to remove heat at a rate of 5 kW m^{-2} from a combustible surface and in doing so maintain its temperature at 350 °C which, for this particular material, is too low for significant volatile release. The flow of inert gas thereby eliminates the ignition hazard. Calculate the required convection coefficient.

Solution

Rate of heat gain by the water = rate of heat loss by the gas

$$4180 \text{ J kg}^{-1}\text{K}^{-1} \times 0.2 \text{ kg s}^{-1} \times 65 \text{ K} = m_g \text{ kg s}^{-1} \times 1100 \text{ J kg}^{-1}\text{K}^{-1} \times 1050 \text{ K}$$

where m_g is the mass flow rate of the gas

$$\Rightarrow m_g = 0.05 \text{ kg s}^{-1}$$

($\approx 2.5 \text{ m}^3 \text{ minute}^{-1}$, at room temperature and 1 bar pressure, of gas of the composition specified. Rate of heat transfer 5 kW to the nearest kW.)

Convection coefficient h calculable from:

$$q' = h\,(T_s - T_g)$$

where q' = heat removal rate per unit area, subscript s denotes the surface and subscript g denotes the bulk gas (i.e. the gas beyond the thermal boundary layer)

$$\Rightarrow h = [5000 / (350 - 50)] = 17 \text{ W m}^{-2} \text{ K}^{-1}$$

8.10b Using the expression:

$$Nu = 0.037 \times Re^{0.8} \times Pr^{1/3}$$

where Nu is the Nusselt number, Re the Reynolds number and Pr the Prandtl number, calculate the speed (u, units m s^{-1}) at which gas will need to be directed at the surface in the previous question if the length of the surface L is 4 m. Use a value of $6 \times 10^{-5} \text{ m}^2 \text{ s}^{-1}$ for the kinematic viscosity ν of the gas, a value of 0.05 $\text{W m}^{-1}\text{K}^{-1}$ for its thermal conductivity k at the film temperature and a value of 0.7 for Pr.

Solution

$Nu = hL / k = 17 \times 4/0.05 = 1360$

$$1360 = 0.037\, Re^{0.8} Pr^{1/3} \Rightarrow Re = 5.9 \times 10^5 = uL / \nu \Rightarrow u = 9 \text{ m s}^{-1}$$

Note

The convection correlation, valid for Re values between 5×10^5 and 1×10^7, can be found in Cengel and Turner (2001).

8.11 A cylinder 10 m long with a diameter of 10 cm forms part of an illuminated display at an entertainment centre. There is an electrical malfunction as a result of which the cylinder reaches a surface temperature of 400 °C, releasing heat at a total rate of 10 kW. Carbon dioxide at 30 °C is available as an extinguishing agent, and is directed at the cylinder in order for there to be no acceleration of heat release for as long as it takes to isolate the malfunctioning part of the display from the mains power. At what speed must the carbon dioxide be directed past the cylinder? Use the correlation:

$$Nu = 0.683 \times Re^{0.466} \times Pr^{1/3}$$

with the following values for the properties of the gas:

$Pr = 0.75$; $\nu = 1.5 \times 10^{-5}$ m^2 s^{-1}; $k = 0.02$ W m^{-1}K^{-1}

Solution

Employ the heat balance equation for the cylinder, using symbol h for convection coefficient as before:

$$10\,000 \text{ W} = h \text{ W m}^{-2}\text{K}^{-1} \times (2\pi \times 0.05 \times 10) \text{ m}^2 \times 370 \text{ K} \Rightarrow h = 9 \text{ W m}^{-2}\text{K}^{-1}$$

$$Nu = (9 \times 0.05/0.02) = 22.5 \Rightarrow Re = 2460 = ud\,/\,\nu \text{ (symbols defined previously)}$$

$$\Rightarrow u = 0.37 \text{ m s}^{-1}$$

Chapter 9:
Evacuation of persons in a fire

In this area of fire protection engineering more than in others, empiricism is evident in the equations used and the codes followed. Proulx (2002) provides a good coverage of the topic, which has been drawn on in the following Exercises.

9.1a A social gathering is taking place in an upstairs room of a hotel. The room is 20 m × 10 m and there are 300 people present. When the fire alarm is raised, exit is by a stair of width 1.3 m. How long will it take for all of the persons to evacuate on the basis of the equation (Proulx, 2002):

flow of persons (persons per second per metre of stair width)

$$= 1.26\,\delta - 0.33\,\delta^2$$

where δ is the density of occupancy (persons per m^2 of floor space).

Solution

$\delta = 300/200 = 1.5\ m^{-2}$

Flow of persons = 1.15 persons per second per metre of stair width

Since the stair is 1.3 m wide, actual evacuation rate = 1.15 × 1.3 = 1.5 persons per second

$$\text{Total time required} = (300/1.5)\ \text{s} = 200\ \text{s}$$

Note

Such factors as time to flashover and rate of fire propagation can however be correlated with the rate of movement of a typical individual person in fire protection engineering and this point will be touched on in a subsequent question.

9.1b The equation used in the previous question suggests that there will be a density of persons corresponding to a maximum efficiency of evacuation. There is therefore a density of persons below which the rate of movement of persons decreases. This is counterintuitive but is in fact well documented not only in relation to evacuation of buildings but also in relation to movement of pedestrians during rush hours in city centres. Differentiate the equation in Exercise 9.1a to obtain the value of the density of persons below which there is slower evacuation than maximum. Insert the value of δ so obtained back into the original equation to calculate the rate of evacuation corresponding to it.

Solution

Assigning the symbol F to the quantity of the left of the equation:

$$dF / d\delta = 1.26 - 0.66\delta = 0 \Rightarrow \delta = 1.9$$

$d^2F / d\delta^2 = -0.66$, i.e. negative signifying that the turning value will be a maximum. Therefore, the optimum density is about 2 persons per square metre of floor area.

$$F = 1.26\,\delta - 0.33\,\delta^2 \Rightarrow 1.2 \text{ persons per second per metre stair width.}$$

Note

These values are quoted by Proulx (2002), but the simple calculations leading to them are not given in detail. Proulx also specifies that the optimum conditions correspond to a walking speed of about 0.5 m s^{-1} along the stairway, just under half the unrestricted walking speed of a fit individual.

9.2 A school hall is 25 m long and 12 m wide and during daily assembly contains up to 1500 persons. There are two exits from the hall, each leading to a stair by means of which pupils can descend to the playground. The pupils have been trained to make for the nearest exit if the fire alarm sounds during assembly. If total evacuation is required to take not more than 5 minutes how wide will the stairs need to be?

Solution

Using the same equation as from Exercise 9.1a, and considering one exit only:

$$\delta = [750/(25 \times 12)]\ \text{m}^{-2} = 2.5\ \text{m}^{-2}$$

$$\Rightarrow F = 1.09 \text{ persons per second per metre stair width}$$

Now if 750 persons are required to evacuate in 300 s, required rate = 750/300 = 2.5 persons per second,

$$\Rightarrow \text{required stair width is } 2.5/1.09\ \text{m} = 2.3\ \text{m}$$

9.3 The dynamics of building evacuation involve many factors including the average separation of any two evacuees in comparison with the length of a single stride on the part of one of them. Although what might be described as 'volumetric movement' of persons displays a maximum when plotted against density occupancy by persons as we saw above, the propulsion speed of an individual always decreases with density of occupancy.

In the planning and design of a school chemistry lab a hypothetical solvent fire is considered. This will propagate at a speed such that there will be no threat to the safety of occupants of the laboratory provided that each can exit at a speed of not less than 1 m s^{-1}. The laboratory has a floor area 6 m × 4.5 m. What is the maximum number of persons that the lab can accommodate consistently with the specified safety requirement? Use the expression:

$$S = S_o - 0.28\delta$$

(derived from data in Proulx, 2002) where S = speed of movement of a person (m s^{-1}); S_o = speed of movement of a person in the absence of restrictions; δ as previously defined. Use a value of 1.25 m s^{-1} for S_o.

Solution

$$1 = 1.25 - 0.28\delta \Rightarrow \delta = 0.90\ \text{m}^{-2}$$

Maximum number of occupants of the laboratory = 6 × 4.5 × 0.90 = 24 persons

9.4 Return to the correlation in the previous question for speed of human movement against density of occupancy of available space. At what density do persons become trapped (i.e. unable to evacuate) according to this correlation?

Solution

Returning to the equation: $S = S_o - 0.28\delta$

If the occupants are 'trapped' $S = 0$

$$\Rightarrow \delta = 1.25/0.28 = 4.5 \text{ m}^{-2}$$

Note

This is known to be about the value of δ at which a 'standstill' situation occurs.

9.5 Return now to the equation considered in Exercise 9.1:

$$F = 1.26\delta - 0.33\,\delta^2$$

for exit of persons per second per metre of stair width F as a function of the density of occupancy. At what value of δ does this predict that persons will be unable to evacuate?

Solution

Inability to evacuate implies $F = 0 \Rightarrow \delta = 3.8 \text{ m}^{-2}$

Note

The equations used in Exercises 9.4 and 9.5 have different bases, one being concerned with speed of human movement and the other with flow of persons. That they give about the same value for the critical density of occupancy above which persons are trapped is significant.

Chapter 10:
Detonations

Introduction

The term 'detonate' has become broadened in common parlance so that it is synonymous with 'ignite' or 'flame', but a detonation is actually a distinct phenomenon from flaming being characterised by supersonic propagation and very high overpressures. Detonations in the chemical industry are rare. Where detonations do occur they tend to be in enclosed spaces or under compression. In the laboratory, detonations can be produced by rapid compression of a gaseous fuel-oxidant mixture. Detonations have a complex structure and are composed of 'cells'. Just as flaming may be prevented if the space available to a flammable mixture is too narrow — the 'quenching distance' effect — so detonations can be prevented from developing under conditions of close containment. What follows in this Chapter has drawn on the coverage by Zalosh (2002).

Cell widths for detonation behaviour in stoichiometric hydrocarbon–air mixtures are reliably known and range from 28 cm for methane–air to 1 cm for acetylene–air. A general rule is that detonation will not occur if the pipe or channel containing the hydrocarbon–air mixture has a diameter of less than one third of the cell width. These ideas are investigated in the Exercise below.

> 10.1a Natural gas (assume to be pure methane) is passed along a 5-inch (inner diameter) pipe. There is air leakage upstream of the pipe with the result that a stoichiometric methane–air mixture is present along part of its length. Predict whether, if there is ignition of this mixture, there will be detonation.

Solution

5 inch = 5 × 2.54 cm = 12.7 cm

Cell width for methane air = 28 cm requiring 28/3 = 9.3 cm for detonation.

The diameter is therefore about 35 % higher than the minimum for detonation, which could result from air leakage.

10.1b Another design rule appertaining to detonations is that a detonation within a pipe, such as considered above, will only propagate beyond the end of the pipe if the pipe length is less than 13 cell widths. On this basis, what is the minimum required pipe length to prevent detonation propagation beyond the pipe in the above example?

Solution

$$\text{Minimum length} = 13 \times 28 \text{ cm} = 3.6 \text{ m}$$

10.1c The impulse on the pipe wall is given by Zalosh (2002):

$I = 0.35\ \rho\, M c_g\, L\, /\, (\gamma + 1)$ (units of pascal seconds)

where ρ = density of the gas before ignition (kg m^{-3}), M = Mach number of the detonation, L = pipe length, γ = ratio of principal specific heats for the burnt gas, c_g = speed of sound in the unburnt gas. For a 5-inch pipe containing methane as previously, find the impulse.

Use the following values: $M = 5.17$, $\gamma = 1.2$, $c_g = 330$ m s^{-1} and $L = 5$ m.

Make your own estimate of ρ if the initial total pressure is 1 bar and the temperature 300 K.

Solution

$$CH_4 + 2O_2 + 7.52N_2 \rightarrow CO_2 + 2H_2O + 7.52N_2$$

The initial mixture contains:

1 molar unit methane + 2 molar units oxygen + 7.52 molar units nitrogen = 10.52 mol

Per mol of total gas:

(1/10.52) = 0.095 mol methane ≡ 1.52 g

(2/10.52) = 0.190 g oxygen ≡ 6.08 g

(7.52/10.52) = 0.714 g nitrogen ≡ 19.99 g

$$\Rightarrow 27.6 \text{ g mol}^{-1}$$

1 m^3 at the initial temperature and pressure contains ≈ 40 mol

$$\rho = 1.10 \text{ kg m}^{-3}$$

$$\Rightarrow I = 0.35\ \rho\, M\, c_g\, L\, /\, (\gamma + 1) = 1500 \text{ Pa s}$$

10.1d Continue Exercise 10.1c to determine the pressure at the walls. Compare your answer with the design stress of mild steel.

Solution

Duration of the detonation = 5 / (5.17 × 330) = 2.9×10^{-3} s

$$\text{Pressure} = \text{impulse/duration} = (1500 / 2.9 \times 10^{-3}) = 0.5 \text{ MPa}$$

The design stress of mild steel is in excess of 100 MPa. This provides a basis for determining how thick the wall needs to be to withstand the pressure. We can see intuitively, however, that a suitable mild steel pipe would withstand this particular detonation. Because of the relief due to the pipes being open, the pressure would not rise to very high values.

10.2 An ethylene oxide/oxidant gaseous mixture detonates with a pressure maximum of about 20 bar. If such a detonation were to occur in a spherical mild steel container of diameter 2.5 m, what thickness of walls is required to prevent failure? Use the expression:

$$e = PD / (4f - 1.2P)$$

where e (m) = wall thickness, D = sphere diameter (m), P is the pressure to be contained (MPa) and f (MPa) the design stress.

Use a value of 135 MPa for design stress of the steel.

Solution

If we take the detonation pressure to be exactly 20 bar, the pressure the wall has to withstand is that value minus atmospheric i.e. 19 bar ≡ 1.9 MPa

$$\Rightarrow e = 0.009 \text{ m (9 mm)}$$

Note

This is of the order of wall thickness that would be likely to be used in the storage of a substance such as LPG with a very high vapour pressure, so is not an inordinately high value. However, a full treatment of this problem would also need to take account of the very high propagation rate and its possible effect on the containing structure.

Ethylene oxide (C_2H_4O) detonation is more powerful than that of many other organic compounds: methane seldom exceeds 15 bar (depending on conditions such as fuel: oxidant ratio). There is a semi-quantitative link between detonation pressure and overpressure of the same hydrocarbon when deflagrating under confined conditions. In fact it has been known for at least 50 years that ethylene oxide explosions often display a large overpressure even when propagation is subsonic. The compound was used to rid houses of insects at one time, such endeavours often resulting in partial demolition of the house! More recent applications of ethylene oxide have been as a non-aqueous sterilising agent.

10.3. Refer to question 10.1d in this Chapter. What wall thickness would a mild steel sphere of diameter 2.5 m need to have to contain the calculated overpressure of 0.5 MPa?

Solution

Pressure to be withstood = (0.5 – 0.1) MPa = 0.4 MPa

$\Rightarrow$ e = 2 mm

Note

The container diameter in the above two Exercises is 2–3 orders of magnitude higher than detonation cell widths.

Chapter 11:
Household fires

11.1 Babrauskas (2003) gives an equation for the peak heat-release rate per unit weight of an ignited Christmas tree as a function of the pre-ignition moisture content:

$$q = e^{5.84 - 0.017M}$$

where q = heat-release rate (W kg^{-1}) and M = moisture content (%).

The equation is based on calorimetric measurements. M varies with time as the humidity of the room changes naturally, and there is in fact hysteresis in the moisture absorption–desorption cycle of woody materials (important in forest fire control).

A Christmas tree as delivered to a private residence has a moisture content of 50 %. By how much will that need to decrease for the heat-release rate in the event of ignition to double?

Solution

Taking logs of the above equation and differentiating,

$\ln q = 5.58 - 0.017M$

$d(\ln q) / dM = -0.017$

$$\ln 2 = -0.017 (M - 50)$$

where M is the moisture content leading to a doubling of the heat-release rate

$$\Rightarrow M = 9\%$$

Note

Could the moisture content descend to such a low value? Perhaps it could if the metabolic functions of the tree were no longer taking place. Leaf tissue tends to continue some functions after the woody part of the structure has become moribund and, given that Christmas trees by definition have a usage time of 12 days, this might provide for safety. Further comments on Christmas tree safety can be found in Jones (2000a).

> 11.2 If a Christmas tree weighs 15 kg and is put up in a hallway with a floor area of 1 m × 4 m, how much will it add to the fire load of the hallway? Express your answer in Imperial units.

Solution

If the moisture content is as high as 50 % the heat value will be approximately 10 MJ kg^{-1}. This will have to be scaled, being significantly lower than that of the benchmark material. The effective weight is then:

15 kg × (10/17) × 2.205 lb/kg = 19 lb

4 m^2 ≡ 43 ft^2

i.e. the contribution of the Christmas tree to the fire load is 19/43 lb ft^{-2}
= 0.4 lb ft^{-2}

Note

We saw in Chapter 1 that residential premises typically have fire loads of 7–8 lb ft^{-2} so the Christmas tree itself does not pose a hazard. The hazard is in its ignitability, exacerbated by the presence of electrically illuminated adornments to the tree which might provide an ignition source if malfunctioning. This point is examined in the following Exercise.

> 11.3 As explained by Babrauskas (2003) a common malfunction in Christmas tree lights leading to an ignition hazard is a section of the wiring between two lamps having a high resistance which, for that reason, can act as a heating element. Such an 'element' need only be releasing heat at about 1 W in order to ignite a flammable surface with which it is in direct contact. What resistance, if receiving 1 amp, would release heat at this rate?

Solution

From Power = Current^2 × Resistance,

$$1\ W = 1^2 \times R \text{ where } R \text{ (ohm) is the resistance}$$

$$R = 1 \text{ ohm.}$$

Note

An unskilled DIY repair to the set of Christmas tree lights could produce an unsatisfactory connection with a resistance of this magnitude. Note that the entire resistance of the circuit would only be of the order of ohms. The hazard is in the concentration of a resistance of the order of an ohm over the very short length of the circuit comprising the connection. We return to this point when discussing fires originating in television sets.

11.4 If a single Christmas tree light produces 0.4 W of power, at what rate is it releasing heat?

Solution

The electricity supply raises the temperature of the filament inside the Christmas tree light and this radiates accordingly, some of the radiation being in the visible range. The decorative cover around the light source probably only lets through a proportion of the visible range so that the Christmas tree lights appear, for example, red or blue. That part of the visible range to which the decorative cover is opaque will be absorbed as thermal radiation inside the cover, either all at once or after multiple attenuating reflections. The decorative cover will almost certainly be opaque to some of the radiation outside the visible range, which similarly will be absorbed within the cover. Any radiation passing through the cover to the surroundings will similarly be absorbed as thermal radiation on encountering an emissive surface such as a wall, floor or item of furniture, possibly after multiple reflections.

In other words, all of the radiation is converted to heat and the rate of heat release is 0.4 W.

Note

It is therefore not helpful, at least when considering fire hazards, to try to resolve the emission from the filament into heat and light. A proportion of it is light – radiation in the visible region, most likely filtered to give a particular colour – along its journey to an emissive surface but all the emitted radiation will eventually be converted to heat.

Note that it makes no difference at all to the reasoning above whether the filament is a black body or a grey body (or, for that matter, a non-grey body though this is unlikely in the case of an electrical heating filament). The emissivity of the filament determines the rate of radiative energy release but is irrelevant to the distribution of the radiation once released.

11.5 An electric iron at its maximum operating temperature of 215 °C is used on a thick blanket made of a synthetic fibre. The user of the iron is called away and does not remove the iron with the result that its hot surface remains in contact with just one equivalent area of the blanket. The material from the blanket is made will start to release flammable and toxic gases when a temperature of 150 °C is reached 2.5 mm below its surface.

How long will this take? Use a value of 1.3×10^{-6} m^2 s^{-1} for the thermal diffusivity α of the blanket material.

Solution

The thickness of the blanket is much smaller than either of its other dimensions so we take conduction to be entirely within the thin dimension. The model that applies is therefore the semi-infinite solid with a step change (from say 30 °C) in temperature at time zero. The solution (Holman, 2002) is:

$$(T(x, t) - T_s) / (T_i - T_s) = \mathrm{erf}\,[x / \sqrt{(\alpha t)}]$$

where $T(x, t)$ = temperature at depth x and time t; T_s = surface temperature (= 215 °C); T_i = initial temperature (= 30 °C). Now $T(x, t)$ = 150 °C in the example being considered, therefore:

$$(T(x, t) - T_s) / (T_i - T_s) = (150 - 215) / (30 - 215) = 0.351$$

$$x / \sqrt{(\alpha t)} = \mathrm{erf}^{-1}\,(0.351) = 0.32$$

Putting $x = 2.5 \times 10^{-3}$ m together with our value for the thermal diffusivity

$$\Rightarrow t = 47 \text{ s}$$

i.e. Danger begins less than one minute after desertion!

11.6a Even after an iron has been disconnected from the electricity supply it is still a hazard for as long as its surface remains hot. Consider an iron, the heated surface of which approximates to an isosceles triangle with base 20 cm and the other lengths 30 cm. Such an iron at an initial temperature of 200 °C is placed with its hot surface vertical in a room at 25 °C and allowed to cool by natural convection. What will the convection coefficient be?

Solution

We first need to apply a little elementary geometry to work out the height between the centre of the base of the triangle to which the iron has been approximated and the apex of the triangle. This dimension, symbol x, is the vertical dimension for convection purposes and is clearly:

$$(30^2 - 10^2)^{0.5} \text{ cm} = 28 \text{ cm}$$

The Grashof number Gr is calculated as:

$Gr = (g\beta\,(T_s - T_\infty)\,x^3)\,/\,\nu^2$
where g is the acceleration due to gravity (9.81 m s^{-2}) and β (K^{-1}) the volume coefficient of expansion. T_s and T_∞ denote respectively the surface and surrounding air temperatures. The product $Gr \times Pr$ is the Rayleigh number, Ra.

In the case of the cooling iron the film temperature is:

$$(200 + 25)\,/\,2 = 112.5\ °\text{C},\ 385.5\ \text{K}$$

and for a gas β is simply the reciprocal of this, 2.6×10^{-3} K^{-1}. The kinematic viscosity of air at the film temperature, from tables, is 2×10^{-5} m^2 s^{-1} and the Prandtl number 0.7, from which:

$$Ra = 1.7 \times 10^8$$

A suitable correlation is then (Holman, 2002):

$$Nu = 0.59\ Ra^{1/4} \text{ where } Nu \text{ is the Nusselt number} \Rightarrow Nu = 67 = hx\,/\,k_f$$

where k_f is the thermal conductivity of air at the film temperature, obtainable from tables as 0.03 W m^{-1} K^{-1}

$$\Rightarrow h = 7\ \text{W m}^{-2}\text{K}^{-1}$$

Note

This is a value typical of natural convection involving a gas. Correlations such as that used have about a 20 % uncertainty on them, possibly more.

11.6b At what rate would the iron in Exercise 11.6a be initially losing heat?

Solution

Setting this rate of heat loss q,

$$q = 7 \times (0.5 \times 0.2 \times 0.28) \times 175 = 34\ \text{W}$$

11.7 It is more likely that an iron being allowed to cool would not be perfectly vertical but would be tilted at an angle θ from the vertical ($\theta = 0°$ being completely vertical and $\theta = 90°$ being horizontal).

Though it has not met with totally uncritical acceptance, an approximate way of correcting convection calculations for such tilt is to insert g cos θ in place of g in the expression for *Ra*. Repeat Exercise 11.7a for $\theta = 15$ degrees.

Solution

$$1.7 \times 10^8 \cos(15 \text{ degrees}) = 1.6 \times 10^8 = Ra \text{ for the tilted iron}$$

$$\Rightarrow Nu = 66$$

We can stop here! The difference between *Nu* values of 67 and 66 is of no possible meaning given the uncertainties in the correlations previously referred to. To treat the iron as being vertical when in fact it has such a tilt is a totally adequate simplification. However, such a simplification would not necessarily be valid in flow regimes with different *Ra* numbers, including liquid flow.

11.8a Television fires are a frequent occurrence and one common cause is bad electrical connections leading to a 'heating element', a point discussed in Exercise 11.3.

Babrauskas (2003) records that if circuit components, including transformers, become hot due to malfunction they can cause increases in surface temperatures up to 500 °C although such temperatures may not be sustained. Lower temperatures originating in the same way would be more likely to become steady.

Consider a component approximating to a cylinder of 10 cm length and 5 cm diameter. If the surface approximated to being 'black', how hot would it have to be to radiate heat at the same rate as a small candle i.e. about 50 W?

Solution

$$50 \text{ W} = (5.7 \times 10^{-8} \times 2\pi \times 0.025 \times 0.1)\, T^4 \Rightarrow T = 486 \text{ K } (213\ °\text{C})$$

Note

A temperature of just over 200 °C could be sustained for long periods if there was no intervention, leading to ignition of the TV casing. This may of course be wood or plastic. If it was plastic, it may or may not be fire-retarded.

11.8b The fact that the malfunctioning component in the previous question radiates at a rate numerically equivalent to the heat-release rate of a candle does not mean, of course, that its action is equivalent to holding a candle to the case of the TV set! One could do a fairly straightforward calculation in which radiation from the hot object to the inside of the TV casing was analysed according to view factors but this would assume transparency to thermal radiation of the entire space between the heat source and the wall which would be a significant departure from reality. Let us assume for illustrative purposes that just one-tenth — 5 W — of the heat actually reaches the inside wall of the TV set casing. This is likely to be a high estimate, especially if we also neglect heat losses through the casing to the outside.

Suppose that the casing is wood and weighs 3 kg. Wood releases flammable volatiles at temperatures as low as 75 °C. If the casing is initially (before the malfunctioning component reaches its final temperature) at 40 °C how long will it be before the inside of the casing is releasing flammable gases and vapours? Use a value of 2000 J $kg^{-1}K^{-1}$ for the specific heat.

Solution

Heat required = 2000 × 3 × 35 = 210 kJ

Time required at a supply rate of 5 W = 12 hours

Note

Time of the order of an overnight period is sufficient even for a component at a temperature as low as 200 °C to bring a TV set close to ignition conditions.

Chapter 12: Detector systems

Introduction

The conventional definition of the time constant τ (seconds) for a heat detector is the same as that for the time constant of a thermocouple (Jones, 2000b). It leads to the equation:

$$dT_d/dt = (T_g - T_d)/\tau$$

where T_g is the temperature of gas impinging on the detector, T_d is the detector surface temperature and t is the time.

12.1 Using the equation above obtain an expression for the temperature difference between gas and detector when $t = \tau$.

Solution

Rearrange and integrate the above equation:

$$\int dT_d / (T_g - T_d) = (1/\tau) \int dt$$

$$\Rightarrow -\ln (T_g - T_d) = t / \tau + \text{constant}$$

At $t = 0$, $T_d = T_o$, the pre-fire temperature, therefore:

$$t/\tau = -\ln(T_g - T_d) + \ln(T_g - T_o) = -\ln[(T_g - T_d)/(T_g - T_o)]$$

$$\Rightarrow \ln [(T_g - T_d) / (T_g - T_o)] = -t/\tau$$

Putting $t / \tau = 1$ gives:

$$(T_g - T_d) / (T_g - T_o) = 1/e = 0.368$$

Note

The time constant is the time at which the temperature difference between hot gas and detector is a fraction 1/e of the difference between the temperature of the gas on detector impingement and the temperature of the room before the fire began.

12.2a By assuming the detector has an increasing temperature due to the gas, spatially uniform at any one time, the time constant can be shown to have the form:

$$\tau = mc / hA$$

where m is the mass of the heat-sensitive part of the detector, c is the heat capacity of the detector material, h the convection coefficient and A the detector area.

What is the time constant for a detector with a thermal capacity mc of 8 J K^{-1}, an area of 0.01 m^2 receiving gas with a convection coefficient 5 W m^{-2}K^{-1}?

Solution

$$\tau = 8 / (5 \times 0.01) = 160 \text{ s}$$

Note

A supplier of heat detectors cannot specify or guarantee a time constant; it depends on the convection conditions. With thermocouples, the time constant is measured by rapid immersion of the tip into boiling water. Such an approach would not be suitable for heat detectors although there is a procedure originating at Factory Mutual in which a detector is plunged into a hot gas to give a measure of its response characteristics.

12.2b The detector in the above Exercise is programmed to respond when the surface temperature reaches 70 °C by sending an electronic message to a fire control centre. With the convection conditions specified above, how long will this take if the gas is at a temperature of 350 °C on reaching the detector? Express your answer in seconds and also as a percentage of the time constant. Let the pre-fire temperature be 30 °C.

Solution

Using the integrated form of the equation for convection to the detector:

$$\ln [(T_g - T_d) / (T_g - T_o)] = - t / \tau$$

With T_g = 350 °C, T_d = 70 °C and T_o = 30 °C, and recalling that τ = 160 s,

$$t = 21 \text{ s (13\% of the time constant)}$$

Note

Calculations of the type seen so far in this Chapter can also be used for sprinkler heads. The Factory Mutual test previously referred to is for sprinkler heads as well as for heat detectors.

> 12.3a A heat detector is placed on the ceiling of a stone cutting workshop in which a tubular fluorescent light is installed. Housekeeping at the workshop is poor with the result that over time the upper surface of the fluorescent light, having an area A_1 of about 0.2 m^2, has become covered with particle debris from the cutting operation. As a result the upper surface becomes heated to T_{fl} = 80 °C when the light is in use for long periods. This radiates to a nearby heat detector and in turn heat is removed from the detector by natural convection with a coefficient h of 4 W m^{-2}K^{-1}.
>
> Treating the lighting assembly surface and the detector as black bodies what temperature will the detector reach during steady operation of the fluorescent light? The view factor F between lighting assembly and detector is 0.01. Take the thermally responsive part of the detector to be a flat disc of 15 cm diameter and the temperature of the surrounding air to be 27 °C.

Solution

For the detector surface under steady conditions, heat received by radiation = heat lost by convection, or

$$5.7 \times 10^{-8} \, F A_1 \, (T_{fl}^4 - T_d^4) = h \, (\pi \times 0.075^2) \, (T_d - 300)$$

Simplifying,

$$1.61 \times 10^{-9}(T_{fl}^4 - T_d^4) = (T_d - 300)$$

Simplifying further:

$$325 = 1.61 \times 10^{-9} \, T_d^4 + T_d$$

Solving by iteration,

$$T_d = 310 \text{ K (37 °C)}$$

Note

There are two philosophically conflicting ways of interpreting the above calculation. One is that the hot surface of the light will not cause activation of an alarm set to respond at, say, 70 °C. Needless operation of alarms is at least inconvenient and possibly expensive. On the other hand, the workshop safety management needs to be aware of a surface at 80 °C. The stone cutting debris will not of course ignite, but something cellulosic or carbonaceous depositing on a surface at that temperature might. Moreover, if the remedial measures of cleaning the light and removing the deposit are not taken, the situation will get more dangerous. If restricted in convective heat loss, fluorescent lights can reach temperatures much higher than 80 °C.

> 12.3b What surface temperature would the fluorescent light be required to reach if the alarm was set to be activated at 70 °C?

Solution

From the above, $1.61 \times 10^{-9} (T_{fl}^4 - T_d^4) = (T_d - 300)$

Setting $T_d = 343$ K,

$$T_{fl} = 449 \text{ K } (176\ ^\circ\text{C})$$

> 12.4a Radiant energy fire detection systems work in one of three wavelength bands (Shifiliti, 2002):
>
> ultra-violet (UV) 0.1 to 0.35 μm; visible 0.35 to 0.75 μm; infra-red (IR) 0.75 to 220 μm.
>
> Select the most suitable type for fire in a smouldering carpet in which the temperature is about 600 °C.

Solution

Utilising Wien's displacement law,

$$(\lambda T)_{max} = 2897.8 \ \mu\text{m K}$$

$(\lambda T)_{max}$ being the product of the wavelength and temperature corresponding to the maximum emissive power.

Setting $T = 873$ K gives: $\lambda = 3.3$ μm

i.e. IR would be the most suitable choice.

Note

The extent to which it is essential to select a detector with a wavelength range encompassing $(\lambda T)_{max}$ is considered in the following part of the question.

> 12.4b For a black body radiating at 873 K (i.e. the situation considered in the question above) what proportion of the energy emitted would be within the wavelength range of a UV detector?

Solution

For this we have to turn to tables for energy increments in particular wavelength bands, such as those in Holman (2002).

For a wavelength of 0.35 μm at an emitting temperature of 873 K,

$$\lambda T = 306\ \mu\text{m K}$$

and this is off scale in terms of the tables in Holman (2002)! The proportion of the energy released in the wavelength range of UV light would be less than one-hundred-millionth of the total.

Note

A UV detector would not respond to the carpet fire.

> 12.4c For a pre-flashover fire at 700 °C, what proportion of the emitted radiation would be in the range of a visible detector?

Solution

$0.35 \times 973 = 341$ μm K, off scale again

$0.75 \times 973 = 729$ μm K, about 10^{-5} of the radiation in the range of the detector.

Note

If the fire is approaching flashover, therefore releasing energy at about a megawatt, the energy release in the detectable range is 10 W.

Appendix: True or False

1. The four element — earth, air, fire and water — concept of the natural order was first proposed by Aristotle.

2. Fire loads are expressed in units of kg m^{-2} or lb ft^{-2}.

3. The benchmark material in the calculation of fire loads is dry wood.

4. Petroleum fractions — gasoline, kerosene, diesel — have calorific values of 42 to 45 MJ kg^{-1}.

5. A rule of thumb in fire protection engineering is that flashover occurs at a total heat release rate of 1 MW.

6. In the treatment of room fires by means of thermal diagrams, flashover occurs when the plot of heat release as a function of temperature and that of heat loss as a function of temperature intersect.

7. Flashover temperatures are typically 900 to 1000 K.

8. A post-flashover fire maintains a steady temperature.

9. In a room fire, times to flashover can be extended significantly by use of a wall material with a high thermal response parameter (TRP).

10. In polymer combustion, to treat the ignition condition as being the attainment of a stoichiometric mixture of released monomer in the surrounding air is a reasonable approximation.

11. Heat fluxes are subject to a conservation law.

12. Thermal diffusivity is a synonym for thermal conductivity.

13. The Prandtl number is the ratio of the thermal diffusivity and the diffusion coefficient.

14. The thermal thickness of a body of specified material, size and shape is an invariant hard number.

15. Solar flux is about 1400 W m^{-2} at high sun.

16. A body that appears white cannot, for that reason, be assumed to be a poor absorber of thermal radiation.

17. The combustible components of a typical vehicle account for about 20 % of its total mass.

18. The body shell contributes little if anything to the fire load of a vehicle.

19. The fire load of a saloon car or a station wagon will not be significantly raised if the luggage space is filled with personal belongings such as clothing.

20. A burning vehicle will have a maximum heat release rate in the neighbourhood of 5 MW.

21. The gasoline vapour inside a vehicle fuel tank is always outside the flammable range of concentrations.

22. Petrol leaked from a ruptured tank in a car accident having occurred during daylight hours will, if the weather conditions are such that attenuation of solar flux is not severe, evaporate completely in the order of tens of seconds.

23. In a situation such as that in question (22) the initial patch of leaked gasoline will have a uniform depth.

24. If petrol leaked in a traffic accident does ignite, the abundance of smoke will depend significantly on the nature of the crude oil from which the gasoline was produced.

25. In traditional bomb calorimetry to determine the heating value of solids and liquids, complete combustion is ensured by burning in oxygen rather than in air.

26. In calorimetry which measures heat release rates rather than simply the total amount of heat released, there is often significant incompleteness of combustion.

27. In the cone calorimeter, radiative and convective heat transfer rates are approximately equal.

28. Where there is incomplete combustion of a solid fuel, an exact energy balance can always in principle be achieved by considering carbon monoxide and smoke.

29. The heat of depolymerisation (gasification) of a polymer is about an order of magnitude lower than the heat of combustion.

30. The heat of pyrolysis of a natural substance such as wood is an example of a heat of gasification and as such has the same sign in the thermodynamic convention as heat of gasification.

31. Water is the most common extinguishing agent in fire protection.

32. In the design of centrifugal pumps that provide water for fire fighting, the most influential factor in terms of delivery rate is the impeller rotation rate.

33. In pump design for fire fighting, two homologous pumps are the same in all characteristics apart from impeller rotation rate.

34. In the thermodynamic analysis of a water-conveying pump, kinetic energy effects are significant, however potential energy effects are less so and perhaps negligible.

35. A simple nozzle is an alternative to a pump in supplying water for fire protection purposes.

36. Fibre proteins are a commonly used class of substance in the manufacture of foams for fire fighting.

37. The equal and opposite concept of fires and extinguishing agents introduced in the main text provides an upper bound on extinguisher capability.

38. A single portable CO_2 extinguisher would be expected to be equal and opposite to a fire approaching flashover.

39. The non-thermal damage resulting from a fire is simply water damage.

40. Post-combustion gas, having been depleted of oxygen, can be used without preparation as an extinguishing agent.

41. A plot of the rate of movement of persons evacuating a building against the density of occupancy (persons per unit area of available floor space) would be expected to display a maximum.

42. In evacuation of buildings, the critical density of occupancy above which persons are trapped (net movement in any direction being impossible) is 3 persons per m^2 of floor space.

43. In a hydrocarbon–air mixture detonation will not occur if the pipe or channel containing the mixture has a diameter of less than one third of the cell width.

44. Dead wood displays hysteresis in its moisture absorption/desorption cycle.

45. A Christmas tree placed in the hallway of a house will significantly raise the fire load of the hallway.

46. In a low-power electrical installation such as a set of Christmas tree lights, an unsatisfactory connection e.g. a dry joint resulting from inexpert soldering, is a possible fire hazard.

47. The rate of heat release by an electric light is the total power minus the rate of visible light release.

48. When a household iron is placed with its heated surface vertical, the coefficient of natural convection from the hot surface to the surroundings will be in the neighbourhood of 50 W $m^{-2}K^{-1}$.

49. Even a mildly heat-releasing malfunctioning component of a TV set can ignite the casing of the set in the order of half a day.

50. The time constant of a heat detector depends on the coefficient of heat transfer to the detector.

True or False: Answers

1. False. Aristotle was familiar with the idea and promoted it, and its eventual wide acceptance was due to him. But the idea itself was not his and actually predates him.

2. True

3. False. It is seasoned wood, which will have a moisture content of about 15 %.

4. True

5. True

6. False. Flashover occurs when the two plots referred to contact each other tangentially.

7. True

8. False. The temperature rises for a period after flashover and eventually drops when fuel depletion starts to have an effect.

9. True

10. True

11. False. Energy of course is conserved, but not energy per unit area, which is flux.

12. False. The definition of thermal diffusivity has the thermal conductivity as the numerator, and this is divided by the product of the heat capacity and the density.

13. False. The Prandtl number is the ratio of the kinematic viscosity and the thermal diffusivity. The ratio in the question – that of the thermal diffusivity and the diffusion coefficient – is the Lewis number.

14. False. The thermal thickness depends on the flux the body is receiving.

15. True

16. True. If the body looks white it must be reflecting in the visible region but the entire wavelength range of thermal radiation extends well beyond this in each direction. The white appearance of the body signifies nothing in terms of its ability to absorb radiation outside the visible range and a white surface can actually be quite a good approximation to a black body. This is not a scientific paradox, simply an accident of terminology.

17. True

18. True. Temperatures are too low for oxidation of the metal to a major extent.

19. False. It was shown in Exercise 5.1c that a saloon car with its boot even lightly occupied by clothing has a significantly higher fire load than a car with an empty boot. This must apply even more to station wagons.

20. True

21. False

22. True

23. False

24. True. In particular, if the crude was aromatic base there will be a high smoke yield.

25. True

26. True

27. False. Radiative dominates and very often convective is negligible.

28. False. As noted in Exercise 7.3 on polystyrene combustion, the residual solid will almost certainly not have the same composition as the original fuel, having been pyrolysed by the heat. This also would have to be noted in an exact energy balance though the necessary information would be unlikely to be available.

29. True. We might also expect the heat of depolymerisation to have a different sign in the thermodynamic sense to the heat of combustion. It is well known, for example, that the manufacture of PVC from vinyl chloride monomer is exothermic, possibly requiring heat removal to prevent thermal runaway. The reverse process will be endothermic.

30. False. The monomer corresponding to wood is glucose, of which cellulose is composed. However, when wood is heated it does not simply depolymerise to glucose but gives a miscellany of products including methanol. There are many concurrent reactions each with its own enthalpy change and no basis for declaring that for the total process:

$$\text{wood} \rightarrow \text{degradation products}$$

will in general be exothermic or endothermic.

31. True

32. False. The most influential factor is the impeller blade diameter.

33. True

34. True

35. False. A simple diffuser, which is a nozzle working in reverse, is such an alternative.

36. True

37. True

38. False. Such an extinguisher would be equal and opposite only to a fire releasing at about a fifth the flashover value of 1 MW.

39. False. The term also includes damage due to smoke deposition.

40. False. The difficulty is in the words 'without preparation'. Such gas must be cooled by heat exchange before being applied to fire protection.

41. True

42. False. The critical value is significantly higher than this, 4–5 persons per m^2 of floor space.

43. True

44. True

45. False. We saw in Exercise 11.2 that a Christmas tree will add about 0.4 lb ft^{-2} to the fire load of such a hallway. The fire load in the absence of the Christmas tree, attributable to furnishings, floor coverings and (possibly) wall panelling, is likely to be in excess of 7 lb ft^{-2}.

46. True. It is likely to have a higher resistance than an equivalent length of conductor elsewhere in the circuit, thereby creating a heating element.

47. False. The visible light is in the wavelength range encompassed by 'thermal radiation' and will itself finish up as heat on striking an absorptive surface, which sooner or later it must. So the rate of heat release is the total wattage of the light.

48. False. The coefficient will be about an order of magnitude lower than this. With air as fluid, a coefficient as high as that in the question could only be obtained if there was vigorous forced convection.

49. True

50. True

References

Babrauskas V. (2002) *Heat Release Rates in Fires.* Handbook of Fire Protection Engineering, Third Edition, Society for Fire Protection Engineering.

Babrauskas V. (2003) *Ignition Handbook.* Fire Science Publishers and Society for Fire Protection Engineers.

Cengel Y. A. and Turner R. H. (2001) *Fundamentals of Thermal-Fluid sciences.* McGraw-Hill, New York.

Crowe C. T., Roberson J. A. and Elger D. F. (2001) *Engineering Fluid Mechanics.* Wiley, New York.

Geankoplis C. J. (1993) *Transport Processes and Unit Operations.* Second Edition. Prentice-Hall.

Gottuck D. T. and White D. A. (2002) *Liquid Fuel Fires.* Handbook of Fire Protection Engineering, Third Edition, Society for Fire Protection Engineering.

Holman J. P. (2002) *Heat Transfer,* Nineth Edition. McGraw-Hill, New York (or any other edition).

Jones J. C. (1993) *Combustion Science: Principles and Practice.* Millennium Books, Sydney.

Jones J. C. (2000a) Spraying Christmas trees. *National Fire Protection Association Journal.* **94** pp 8–9

Jones J. C. (2000b) *The Principles of Thermal Sciences and their Application to Engineering.* Whittles Publishing, Caithness and CRC Press, Boca Raton. pp 152

Jones J. C. (2003a) *Hydrocarbon Process Safety: A Text for Students and Professionals.* Whittles Publishing, Caithness. US Edition published by Pennwell, Oklahoma. pp 292

Jones J. C. (2003b) Analysis of heat flux data for a cone calorimeter. *International Journal on Engineering Performance-Based Fire Codes.* **5**(18)

Jones J. C. and Wake G. C. (1990) Measured activation energies of ignition of solid materials. *Journal of Chemical Technology and Biotechnology.* **48** pp 209–216

Lie T. T. (2002) *Fire Temperature-Time Relations.* Handbook of Fire Protection Engineering, Third Edition, Society for Fire Protection Engineering.

Proulx G. (2002) *Movement of People: The Evacuation Timing.* Handbook of Fire Protection Engineering, Third Edition, Society for Fire Protection Engineering.

Rogers G. F. C. and Mayhew Y. R. (1980) *Thermodynamic and Transport Properties of Fluids.* Blackwell.

Scheffey J. L. (2002) *Foam Agents and AFFF System Design Considerations.* Handbook of Fire Protection Engineering, Third Edition, Society for Fire Protection Engineering.

Schifiliti R. P. (2002) *Design of Detection Systems.* Handbook of Fire Protection Engineering, Third Edition, Society for Fire Protection Engineering.

Tewarson A. (2002) *Generation of Heat and Chemical Compounds in Fires.* Handbook of Fire Protection Engineering, Third Edition, Society for Fire Protection Engineering.

Zalosh R. (2002) *Explosion Protection.* Handbook of Fire Protection Engineering, Third Edition, Society for Fire Protection Engineering.